ROCKING P
RANCH

THE WEST SERIES
SERIES EDITOR: George Colpitts

ISSN 1922-6519 (Print) ISSN 1925-587X (Online)

This series focuses on creative nonfiction that explores our sense of place in the West – how we define ourselves as Westerners and what impact we have on the world around us. Essays, biographies, memoirs, and insights into Western Canadian life and experience are highlighted.

No. 1   *Looking Back: Canadian Women's Prairie Memoirs and Intersections of Culture, History, and Identity*
S. Leigh Matthews

No. 2   *Catch the Gleam: Mount Royal, From College to University, 1910–2009*
Donald N. Baker

No. 3   *Always an Adventure: An Autobiography*
Hugh A. Dempsey

No. 4   *Promoters, Planters, and Pioneers: The Course and Context of Belgian Settlement in Western Canada*
Cornelius J. Jaenen

No. 5   *Happyland: A History of the "Dirty Thirties" in Saskatchewan, 1914–1937*
Curtis R. McManus

No. 6   *My Name Is Lola*
Lola Rozsa, as told to and written by Susie Sparks

No. 7   *The Cowboy Legend: Owen Wister's Virginian and the Canadian-American Frontier*
John Jennings

No. 8   *Sharon Pollock: First Woman of Canadian Theatre*
Edited by Donna Coates

No. 9   *Finding Directions West: Readings That Locate and Dislocate Western Canada's Past*
Edited by George Colpitts and Heather Devine

No. 10   *Writing Alberta: Building on a Literary Identity*
Edited by George Melnyk and Donna Coates

No. 11   *Ranching Women in Southern Alberta*
Rachel Herbert

No. 12   *Rocking P Ranch and the Second Cattle Frontier in Western Canada*
Clay Chattaway and Warren Elofson

UNIVERSITY OF CALGARY Press

ROCKING P RANCH AND
The Second Cattle Frontier in Western Canada

BY
Clay Chattaway and Warren Elofson

The West Series
ISSN 1922-6519 (Print) ISSN 1925-587X (Online)

© 2019 Clay Chattaway and Warren Elofson

University of Calgary Press
2500 University Drive NW
Calgary, Alberta
Canada T2N 1N4
press.ucalgary.ca

This book is available as an ebook which is licensed under a Creative Commons license. The publisher should be contacted for any commercial use which falls outside the terms of that license.

LIBRARY AND ARCHIVES CANADA CATALOGUING IN PUBLICATION

Chattaway, Clay, author
    Rocking P Ranch and the second cattle frontier in Western Canada / by Clay Chattaway and Warren Elofson.

(The West series, ISSN 1922-6519 ; no. 12)
Includes bibliographical references and index.
Issued in print and electronic formats.
ISBN 978-1-77385-010-8 (softcover).—ISBN 978-1-77385-011-5 (Open Access PDF).—
ISBN 978-1-77385-013-9 (EPUB).—ISBN 978-1-77385-014-6 (Kindle).—
ISBN 978-1-77385-012-2 (PDF)

    1. Rocking P gazette. 2. Macleay family. 3. Ranches—Alberta—History.
4. Ranchers—Alberta—History. 5. Ranching—Alberta—History. I. Elofson, W. M.,
author II. Title. III. Series: West series (Calgary, Alta.) ; 12

FC3670.R3C53 2019   636'.010971234   C2018-905507-3
                                                       C2018-905508-1

The University of Calgary Press acknowledges the support of the Government of Alberta through the Alberta Media Fund for our publications. We acknowledge the financial support of the Government of Canada. We acknowledge the financial support of the Canada Council for the Arts for our publishing program.

This project was funded by the Government of Alberta through the Alberta Historical Resources Foundation.

Copyediting by Peter Enman
Cover credit: *Rocking P Gazette*, February 1924, Cover. Property of the Blades and Chattaway families and their descendants.
Cover design, page design, and typesetting by Melina Cusano

CONTENTS

Preface	VII
Introduction: The Macleay Family and the *Rocking P Gazette*	1
I. THE SECOND CATTLE FRONTIER IN WESTERN CANADA	7
1 Go West Young Men	9
2 The Extended Family Period: Riddle and Macleay Brothers	17
3 Nature's Fury and The Tattered Dream	29
4 The Rocking P Ranch (and Farm)	41
5 Enlisting the Nuclear Family, 1909–1925	57
6 Finance Matters	71
II. THE *ROCKING P GAZETTE*	87
7 Introducing the *Rocking P Gazette*	89
8 The Rural West	97
9 Country Entertainment	121
10 Principles of Need	145
11 From Religion to Race	159
12 Reinforcing Family Values	185
Conclusions	203
Notes	213
Bibliography	241
Index	249

PREFACE

This book is the product of collaboration between a well-known cattle rancher in the Porcupine Hills some ninety kilometres south of Calgary, Alberta, and an academic from the University of Calgary who has also had long experience farming and ranching in the province. The academic is Professor Warren Elofson. The rancher is Clay Chattaway, grandson of Roderick Riddle and Laura Marguerite Macleay, whose story is told here. Mr. Chattaway has donated his own vast knowledge of his family's history and the long-term development of the beef industry in western Canada, and he has granted access to the very valuable Macleay family papers. This collection provides opportunity for a deep analytical assessment of a single ranching operation. It also facilitates our central thesis: that during the twentieth century, the family unit has ultimately been more capable than any other business structure in achieving agricultural sustainability in the northern foothills region of the Great Plains.

INTRODUCTION

The Macleay Family and the *Rocking P Gazette*

Following is the story of the Rocking P ranch, owned and operated by the family of Roderick and Laura Macleay in the foothills of southern Alberta in the early 1920s. The story is based primarily on the *Rocking P Gazette* newspaper, which was produced and edited by the Macleays' two young daughters, Dorothy and Maxine. In conjunction with the rest of the Macleays' personal and business papers, the newspaper provides a great array of insights into the practical, financial, and cultural attributes of this particular type of agricultural unit at a specific time and place in western Canadian history. This is all the more significant because scholars generally have very few really bountiful primary materials with which to chart rural family history. A number of huge company ranches that opened the first cattle frontier in the Canadian and American Wests in the 1880s had a hired manager who was required to report on a regular basis to a head office in the East or overseas and, most importantly, to preserve onion skin copies of his letters and other documents in prescribed letter books. Families did not have the same personnel or requirements, and they tended to be haphazard about what, if any, records they kept. Moreover, they seldom had a way to duplicate any written communications. When a primary source such as this is uncovered, it is a precious find.

It was, moreover, the family ranching operation, not the company outfit, that was destined to establish a sustainable form of agriculture on the northern Great Plains. Major company producers such as the Cochrane ranch, the Bar U, the Walrond, and the Oxley are credited with forming the first cattle frontier in the Canadian West because, starting

1

in 1881, they brought in the initial great herds to graze the natural grasslands in Alberta and Assiniboia (now southern Saskatchewan). However, by the end of the killing winter of 1906/7 they, like their counterparts on the other side of the American border, all failed, largely because they used an open range system that left thousands of their cattle to fend for themselves year round on open range leases of up to 300,000 acres. This approach, known as the "Texas system," had seemed to prove itself for a time in the more moderate environment of the southern United States, but it was patently inappropriate on the northern Great Plains, where it subjected the herds not only to the winter blizzards that regularly strike that region but also to bands of hungry wolves and cattle rustlers and diseases that spread among the animals as they mixed and mingled over countless acres.

We are referring here to the period running approximately from the turn of the twentieth century to World War II during which family ranching became established in the Canadian West as the second frontier, because the word "frontier" speaks of a new beginning, a new way of living, and a new way of doing. Historians have devoted considerable print over the years to agriculturalists' endless search for systems of production suitable to the climate and terrain on the Great Plains. James Gray, Paul Voisey, André Magnan, Donald Worster, Max Foran, Terry Jordan, Courtney White, David Breen, and the present authors have all in one way or another attempted to show how ranchers and farmers adjusted their practices to find methodologies that would enable them to extract a profit (or a living) out of the land. In overly simple terms, the present study contributes to this work by arguing that the family operations were far more successful in the effective transition from a grazing culture, the most elementary agricultural form, to a more complex approach. They employed practices the grazing companies were reluctant to use, such as fencing their pastures to manage their land and livestock; putting up enough feed to see their cattle safely through the longest cold spells; protecting their stock from both two- and four-legged predators, and, ultimately, solving the problem of overgrazing of which the companies had been guilty. A mixed farm evolved by 1910 because grazing by itself had shown its limitations. Family farms were better suited to the environment than companies whose short-term goal was satisfying shareholders. The mixed farming methods they embraced included horse breeding, dairying, pork, poultry and egg production, and even harvesting field crops such as oats, barley, and wheat. The family also provided an onsite labour pool and, as we will see, this enabled the regular blurring of traditional gender roles.

While the Macleay operation was to become in many ways typical of family ranches in the foothills of Alberta, it was in one important respect far from average. As we will see, Roderick Macleay, who started it all, was a very ambitious man. He set out to build a cattle empire, and he had the unqualified support of his wife, Laura. Between the two of them, they managed to accumulate, and eventually to sustain, uncommonly large land and livestock holdings. Theirs was one of a select few family ranches that, though small compared to a number of the earlier company outfits, exceeded the average mark by a considerable amount. By 1914 the average family unit on the prairie had over 300 acres of freehold and about 1,000 acres of rented land on which it grazed two or three hundred head of cattle. The Macleays and others, including the Crosses on the A7 ranch a few miles to the east, the Cartwrights on the D ranch southwest of Longview, the McIntyres on the Milk River Ridge in southeastern Alberta, the McKinnons on the LK ranch near Brooks, and the Copithornes on the CL west of Calgary, all eventually evolved into multi-generational enterprises with much larger land and livestock inventories.

What can be said, however, is that it was employing the conventions of the family unit with more than average energy and determination that eventually made such operations what they were to become. Consequently, the Macleays' story casts valuable light on the emergence and development of the second frontier and on the factors that gave it sustenance. As we will show, this statement refers not just to a system of agricultural production but to culture in the broader sense as a lifestyle. We hope as well that our study will help to fill a gap in the historiography of the West. In part because documentable evidence is short, historians have never been able to provide an in-depth elucidation of family ranching operations in the foothills region to complement, for instance, Paul Voisey's thorough examination of the wheat belt community in and around the town of Vulcan, Alberta, to the east. What follows should make a valuable contribution to that end.

Once we have described the history of the Macleay family, we will turn to our second major objective—an intricate examination of the *Rocking P Gazette* newspaper itself. This will prove a considerably more complex, and potentially rewarding, undertaking than one might on the surface expect. Dorothy Margaret Macleay, aged fourteen, and her younger sister, Gertrude Maxine, aged twelve, started the *Rocking P Gazette* in 1923, edited it, acted as its principal reporters, wrote many of its articles and stories, and sketched and painted nearly all its art. At first glance, therefore, one might expect the paper to be a rather ornate and charming artifact—something out of the past, reflecting a life of school and games and play. On closer examination, however, we see a lot more than that. Over the course of a year and a half, the Macleay sisters directed the seventeen monthly editions of their publication at an audience consisting of their father and mother and all the relatives, cowpunchers, teachers, and cooks who lived and worked on any of the family's extensive holdings. The two girls attempted to make each of their editions as much as possible like regular prairie newspapers. Therefore, though in somewhat different order than in those publications, they included within their pages a "Local News" section featuring the people and events in their community (in their case Macleay ranches), numerous ads for consumer articles that local people relished, humorous tales to which they could relate, jokes and poetry, fiction, and real life stories that reflected and resonated in their cattle ranching world. The result is a multi-faceted representation of daily life in the foothills of western Canada in the 1920s such as we have not been able to find elsewhere.

It needs to be said as well that the paper's standards of scholarship are very high considering the age of its two editors. There are three reasons for this. The first is that Dorothy and Maxine's very capable and motivated teacher, Miss Ethel Watts, lived on the ranch with them and was thus able constantly to oversee their work. She set up each edition of the paper by providing the "index" (or table of contents) at the beginning, and she regularly wrote a short story or poem (or both) for each issue. Evidently, she also vetted all the monthly editions of the paper before they came out. The second factor is the girls' relatively high educational standards, which we will explain in full below. The third is their innate artistic talent, which, as we will also demonstrate, is unmistakable in their written and visual art.

One ingredient the editors of the *Rocking P Gazette* were able to incorporate into their monthly offerings, and that also merits mention at this stage, consisted of regular contributions from numerous of the Macleay ranch hands. A number of the young men who worked on one or more of the family holdings in these years wrote stories or news items for the paper that enable the reader to participate with them in daily activities from working the cattle herds to putting up hay to feeding pigs to engaging in sports such as baseball and rodeo to even less complicated activities, including getting drunk. Most importantly, the men's offerings also enable the reader to follow the long-standing cowboy traditions of composing poetry, compiling country and western songs, and generally participating in the world of rural entertainment. In this we see the second iteration of a tradition that looked back to the first cattle frontier and that would, rather incredibly, continue to grow and to flourish in and well beyond rural communities in both Canada and the United States through to the present.

Ultimately, then, what we expect to achieve overall by first illustrating the history of the Macleay ranchers and, secondly, examining the *Rocking P Gazette* in its many intricacies, is to construct a pathway to a better understanding of family agriculture, and of ranching culture in the broadest sense, during the second cattle frontier on the eastern slopes of the Canadian Rockies. However, what we are saying is we want to do more than that, too. The *Rocking P Gazette* comes from the hearts of people who actually lived their life on the second frontier, experienced its circumstances and conditions first-hand virtually every day, and wrote down their thoughts, not just to entertain but also to inform and enlighten each other about the things that mattered to them in the life they shared. It is an expression of their collective mentality, a compendium of their disparate views on whatever happened to strike them as worthy of attention on a particular occasion. It is raw, genuine, and unabridged, and once the Macleays' western history is firmly in our minds, it enables us to see the second cattle frontier through the eyes of the whole spectrum of individuals and types who played a significant role in making it happen.

Note

We hope that readers will be encouraged to consult the various versions of the newspaper as they work their way through our pages. Because the paper is so large—some 1700 pages of handwriting, paintings, and sketches—it cannot be reproduced in full here. However, the entire collection can be found online at http://contentdm.ucalgary.ca/digital/collection/rpg.

I.
The Second Cattle Frontier in Western Canada

1

Go West Young Men

At the turn of the twentieth century, Roderick Riddle Macleay was a twenty-one-year-old living in the small town of Danville in the Eastern Townships of Quebec. He was still single, and he was suffering from what was considered a life-threatening case of chronic rheumatoid fever. One day he met a family friend and former neighbour, George Emerson, who was on a return visit to his earlier home in Danville from his cattle ranch in the Alberta foothills. Emerson made an indelible impact on the young Macleay as he proceeded to fill Roderick's head with his own experiences on the western frontier. We cannot be sure *exactly* what Emerson said, but we do know that he had been on the frontier from the beginning—that, indeed, he had been one of its most important players and that, therefore, he knew the entire story.

Emerson unquestionably knew that cattle had first appeared in the foothills and mountainous regions of western Montana after making the long journey north from Texas along the Chisholm and the Goodnight-Loving trails. Starting in the 1870s, increasing numbers of these cattle were driven across the border to Alberta to feed Indigenous people facing starvation with the destruction of the bison herds. Emerson himself, and Tom Lynch, who had migrated west from Missouri, had driven in hundreds of horses and cattle to sell to a number of small ranchers, many of them former North West Mounted Police officers. In 1879, the two men had also brought a thousand cattle and horses north to start up their own ranch along the Highwood River west of the town of High River.[1]

It was in the late 1870s and early 1880s that the era of the so-called "great ranches" began and suddenly turned this northward bovine trickle into a fast-running stream as new grazing corporations suddenly

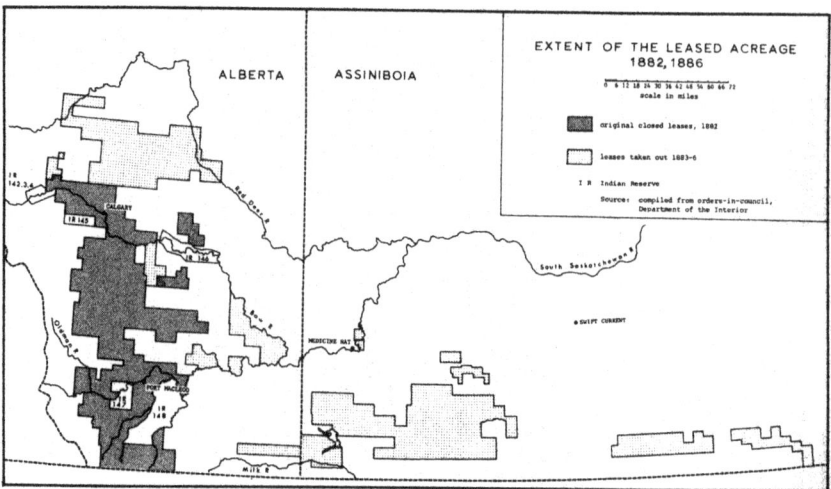

FIGURE 1.1. Extent of the leases in western Canada, 1886. Derived from Orders in Council, Department of the Interior. Simon Evans, "The Passing of a Frontier: Ranching in the Canadian West, 1882-1912." Unpublished MA thesis: University of Calgary, 1976, reprinted in David H. Breen, The Canadian Prairie West and the Ranching Frontier (Toronto, University of Toronto Press, 1983), 46.

became infatuated with the opening frontier. Almost overnight, big joint-stock companies originating in Boston, New York, Edinburgh, London, and Montreal began to invest pools of surplus capital in the western cattle business. Many of them were situated south of the border, but a number, including the four mentioned above in the Introduction (the Cochrane ranch, the Bar U, the Walrond, and the Oxley) established themselves in the foothills directly south and west of Calgary. Others, including the Circle ranch, occupied the hills of the Milk River Ridge along the American border, and still others, including the Canadian Agricultural Coal and Colonization Company, or 76 ranch, the Turkey Track, the Circle Diamond, and the N Bar N, settled in the region running farther east from the Cypress Hills to the Wood Mountain area in Assiniboia Territory.[2] The cattle numbers in western prairie Canada rose in less than a decade from a few thousand to over half a million.[3]

Emerson very likely described the young men who stayed in Canada after trailing in the herds to work on the big company outfits, and who imported the riding, roping, and droving skills of the cowboy with them—men including the famous black cowboy, John Ware; the manager and then part-owner of the Bar U, George Lane; the celebrated bronco buster, Frank Ricks; the one-time foreman of the Bar U, Everett Johnson; the Cochrane ranch cowboys, W. D. Kerfoot, Jim Dunlap, and a Mexican known as CaSous; and the first Walrond ranch foreman, Jim Patterson. All these men had, like Emerson and Lynch, learned their trade on ranches in the American West. They were practised cowhands, and they became the role models for a lot more young men who followed them in from eastern Canada and Great Britain. Many of these men were "wannabes" who, immediately upon stepping off the train in Calgary, headed to local shops to secure the wide-brimmed hat, the boots, the bright shirt and bandana, and the spurs and chaps they needed to look the part of the working cowboy. Some, known widely as "remittance men," had been sent to the frontier with family financial support, mainly to avoid disgracing their parents at home. Many of them failed miserably, turning to drink, prostitutes, and general dissipation. Others, though, signed on with one of the cattle operations and learned to ride, rope, brand, and even handle a six-shooter. Together, they and their American teachers transformed life in the western foothills. In the 1870s, one rancher stated, "no one" on the northern Great Plains had even "heard tell of a cowboy," but by 1883, "leather chaps, wide hats, gay handkerchiefs, clanking silver spurs, and skin fitting high-healed [sic] boots ... had become an institution."[4]

Emerson's stories about life in the West came at a time when Rod Macleay, like many young men from eastern Canada, the eastern United States, and Europe, was being subjected to a great deal of promotional literature advertising the vast potential of cattle ranching on the Great Plains from Texas all the way north to Alberta. One pamphlet the Canadian government had recently published seemed to be lauding the very region that Emerson was talking about. This land "stands unequaled among the cattle countries of the world," it asserted, and "is now looked upon as one of the greatest future supply depots of the British markets. Great herds of cattle roam at will over seemingly boundless pastures." Herds are "now being sent into this country all the way from Ontario to fatten on the nutritious grasses of these western plains and it is reckoned that after

Figure 1.2. Edward L. Wheeler, *Deadwood Dick's Eagles; or, The Pards of Flood Bar* (New York: M.J. Ivers & Co., 1899). Image courtesy of the Northern Illinois University Libraries' *Nickels and Dimes* site, https://dimenovels.lib.niu.edu/.

paying the cost of freight for 2,000 miles the profit will be greater than if those cattle had been fattened by stall feeding in Ontario."[5] Young male easterners had also been inundated in their adolescent years by a host of dime novels eulogizing cowboy types like Stampede Steve, Buffalo Bill Cody, or Kit Carson, who had attained celebrity by performing heroic deeds in the opening West like killing "savages," or saving a damsel in distress from a charging bull, or imposing justice on frontier outlaws.[6] Many easterners had also been influenced by a multiplicity of romantic novels about young males heading West and ultimately achieving great things on a cattle ranch by testing their previously untapped physical and cerebral talents against the stimulating qualities of the western environment.[7] St George Henry Rathborne's *Sunset Ranch*,[8] Owen Wister's *The Virginian*[9] (for which the hero was modelled on Everett Johnson who moved to Alberta in 1888), H. L. Williams' *The Chief of the Cowboys*,[10] and Canadian Ralph Connor's *Sky Pilot*,[11] to name but a few, made a very distinct impression on maturing youngsters and helped eventually to draw many of them to the opening cattle frontiers on both sides of the forty-ninth parallel.

At a time when print was still by far the most significant form of media, images created by these sorts of publications could, and we would argue did, have a much greater impact on the reading public than scholars have previously assumed. Even if Macleay was relatively impervious to the more outlandish representations, he must already have had an upbeat image of what could be achieved on the first cattle frontier when he heard from Emerson. One thing the latter could not have told him, because he would not have known at this time, was that in relatively short order all the company ranches would be gone.[12] However, he did know, and he almost certainly did inform Macleay, that a new type of cattleman was now pouring into the foothills to take up ranching on a much smaller scale. The individuals who were of this type were, like Emerson himself, settling on a quarter section (160 acres) of more or less free land under the Dominion Lands Act of 1872 and then leasing or purchasing further holdings on which to pasture their smaller herds and to put up enough wild hay to sustain their stock through the long winters.[13]

The homesteaders were particularly relevant to Macleay. He was in a sort of holding pattern in his own life. He had been working with his father running a lumber mill in Quebec, but he was also trying, without much success, to make up his mind about what he wanted to do for a living in the longer term. Suddenly, through the homesteaders, he was able to see his way to a new future on land of his own. Shortly after his conversations with Emerson, therefore, he decided he was going to join the "settler revolution" and to fill as much as he could of whatever life he had left working cattle on the western prairies.[14] His exuberance was infectious, and when he announced to his friends that as soon as possible he would be making an exploratory trip to southern Alberta to see firsthand what the country was all about, three other Danvillites decided they wanted to go with him.[15]

A few weeks later, Rod, his cousin Douglas Riddle,[16] Uncle Joseph Riddle,[17] and their mutual friend, Marvin Morrill, climbed aboard a Canadian Pacific Railway train for Calgary. From there, the men took the C & E (Calgary & Edmonton) line some twenty miles south to the town of Okotoks, where they acquired a team and wagon for the final leg of their journey to a ranch on the lower Highwood owned and operated by Joe Riddle's son-in-law, David Thorburn. Thorburn was delighted to see them, and he showed them the western hospitality, which Emerson had told them to expect, immediately offering them a hot meal and accommodation for the night. The next morning, after consuming an ample breakfast of fresh eggs, toast, and coffee, Rod, the younger Riddle, and Morrill left Uncle Joe to visit with his daughter and son-in-law and headed to Emerson's place some thirty-five miles west on Pekisko Creek, a tributary of the Highwood.

During the next few days the three were able to feel they were experiencing the life of true westerners. Rod got his first try at roping livestock from the back of a horse, and he and his companions pitched in and helped put up fresh hay for two other former Danvillites, the McKeage brothers, Billy and John. They also stooked green feed and helped track down some missing horses. They found some time to go hunting and fishing, too. On one outing they shot seventeen sharp-tailed grouse and a goose, and on another they pulled a 24-inch trout from the Highwood River. One day Rod and yet another former Danvillite, Willis Wentworth, went to a big lake to hunt ducks. Rod was thrilled with it all. "We have our larder well filled," he wrote in his diary. "What a contrast between what I am now doing and what I would be doing if I were at home. Willis fired a shot and the noise of the ducks rising up from the water is simply deafening. There are thousands upon thousands congregated on this lake."

For the time being, the western spirit of all three men was and remained high. When they returned to Quebec, Rod, Douglas Riddle, Marvin Morrill, and Rod's brother, Alexander, who had not been able to make the earlier trip, signed a contract to form a ranching partnership. They also pooled their resources (and presumably, accepted financial help from their respective families), to purchase the basic items they felt they would need to get started in the West.[18] When the spring of 1901 arrived, they packed 295 head of cattle, some horses, haying machinery, kitchen utensils, furniture, bedding, clothes, and lumber, onto railway cars and headed back out to the Alberta foothills for a more prolonged stay.

2

The Extended Family Period: Riddle and Macleay Brothers

Once in the high country south and west of Calgary for what they believed would be the long term, the four young men experienced the exhilarating sensation that comes from a sense of rebirth. As they began exploring the hill country around Emerson's place for a location for their new ranch, some of the scenery they encountered plainly added to this. Rod was thrilled when he and Billy McKeage came across a beautiful area nestled among the hills with lots of grass and good water, with a stream meandering through it. He wrote in his diary: "struck on the best location on the whole of Alberta"—a valley with "lots of water and thousands of tons of hay."[1] This was section 32, township 16, range 1 west of the 5th meridian. Rod, Alex, Douglas Riddle, and Marvin Morrill ultimately claimed all of the section, each filing for a homestead on one of the four quarters.[2] Uncle John Riddle then bought section 31-16-1-W5, a C & E Railway section, to provide a neutral location on which the partners would build their company house.[3]

Morrill, who evidently had different long-term plans from the beginning, left the partnership after less than a year. However, for the time being, the three remaining partners all clung to the belief that the future would be bright, and they bought him out. Now known as Riddle and Macleay brothers, they proceeded to acquire more land through lease and purchase.[4] By 1904 they controlled a contiguous block of six sections, nearly 4,000 acres.[5] The partners also invested, mostly in the form of their own labour, in infrastructure at the home place. They started with living quarters for themselves. At first, they slept in a hastily constructed wall tent, which kept them a little too close to nature for comfort. For one two-week period in May 1901 it rained more or less constantly while they drove wagons back and forth daily from the terminus at Cayley, hauling in lumber and supplies from their settler's railway cars. On 10 May, a cold rain turned into two feet of snow covering the hills. Rod had made a trip for supplies and spent the night in town. When he arrived back at the camp he found the others all under the collapsed tent canvas, shivering in the cold. This no doubt ignited their enthusiasm for a permanent structure.

Following their original plan, they soon began using the precut lumber to start their new home on a hillside on the southeast quarter of section 31. They built the basic 24-foot by 36-foot two-storey frame house from sill to shingles in just twelve days (15–27 May). Willis Wentworth, a Danville friend who came from Calgary with his carpentry skills, and the firm in Danville that had originally cut the lumber, made their job much easier than it otherwise might have been. The finished product was an early 1900s version of a prefabricated house. As soon as it was habitable they moved in, happily folding the tent and quickly turning their attentions to other buildings and fences. The poles and lumber for the barn they got from the Findlay Brothers Mill on Timber Ridge, about fifteen miles southeast of the home place. "The Colonel" (Rod's nickname for George Emerson) and Uncle John put up the pole frame structure, mortising the posts and beams in a typical Eastern Township style. On the "fine day" of 6 July, they all pitched in and built a henhouse and then "a House of Parliament," as they called the outdoor toilet. At that point, the home place was more or less complete.

The partners' diet in these days was based largely on beef, and for greens and the all-important potato staple, they planted a vegetable garden on the top of a hill about a mile east of the house. Once that was done they, like most homesteaders in those days, made a start at enclosing their pastures in order to control their roaming livestock. In two weeks, they constructed some six miles of three-strand barbed-wire fence. The work was not easy. Most of the posts, which they drove into the ground at intervals of fifteen paces, they had to cut out of the crooked, tangled mess of indigenous willow bush growing locally, and sharpen them by hand.

FIGURE 2.1. Macleay home place circa 1930. Photograph property of the Blades and Chattaway families and their descendants.

Thereafter, erecting fences around their pastures and constructing corrals at the home place became the main order of business whenever the men were not preoccupied with other matters. Even if the necessary materials had been available, purchasing them was an unthinkable expense. Eventually, to find enough logs for posts and rails, they had to look well beyond their immediate area. They would take a crew of up to four farther west to the coniferous forests nearer the mountains some twenty miles away for a few days, to fell, trim, and section trees. They would bring back what they could and stockpile the rest. Later, one or two of them would go back out whenever possible on a day trip to draw more of the wood home. When the stockpile was gone, they would repeat the process.

In this second frontier period, ranchers needed to add to their supply of posts to enclose their pastures if and as their land base expanded and also to replace existing posts as they were broken by falling trees and as they aged and succumbed to rot. Therefore, the job of cutting the logs out of the bush was, for many, like fencing itself, something they had to attend to on a regular and long-term basis. For Rod Macleay, this was to be the case over and beyond the next two decades. The other task to which he and the others soon turned their attention was haying. Seeing that the tall native grasses—the rough fescue on the uplands and the wheat grass on the lower elevations—were in full bloom and voluminous, they started on 29 July 1901. Working with two hired men, they managed to put up 202 wagonloads by 22 August. From that point on, haying became a major part of the annual cycle, as the feed proved essential. In early 1904, Rod wrote in his diary: "This has been a long, long winter for me. Will be almighty glad when spring sets in." They had not started feeding their cattle until 21 December but then had to do so continuously through to 7 April. "With a 'hurrah,'" Rod wrote, "that was it for the year. Though two inches of snow had fallen the night before, it was all gone by evening and the hills were greening up." Even so, "when April came so did the occasional spring blizzard."

Douglas Riddle's sister, Margaret, came out from the East to help out wherever she could, and she and at least one other woman were on hand during haying in 1904. The crops were some distance from the buildings, so the haying crew set up a camp with a wall tent near a creek. With two mowers, a dump rake, a sweep, and an overshot stacker, a five-man crew was required. The women kept them supplied with baked goods such as bread and pies. There may also have been a camp cook and possibly an extra hand to catch up fresh horses for the midday change and sharpen mower sickles. With a minimum of five teams in the field, there would have been at least twenty horses in camp and more in reserve at the buildings. The crew started on 27 July and, after a short rain delay, hayed straight through the whole month of August. They finished the field work on 7 September. By measuring the length, breadth, and over throw of the haystacks, and applying a standard formula, Rod and his partners calculated they had about 400 tons of hay.

In all probability, the cattle they brought west were Shorthorns. We do know that they were stockers to economize shipping (220 yearlings and 61 two-year-olds were selected with the help of Doctor A. Lyster, a well-known veterinarian), with the plan of selling them when ready and then buying a herd of cows, which they did in 1903. These, naturally, were Shorthorns, as they are a hardy breed and were most popular in this period. The male calves would be castrated and kept for three to five years until finally large and fat enough to be sold on the slaughter

FIGURE 2.2. Mowing Hay. Charlie Waddell, ranch hand, A7 Ranch, Nanton, Alberta., [ca. 1910-1912]. Glenbow Archives, NA-691-26.

FIGURE 2.3. Richard Shore raking hay, with a dump rake, which rakes the hay into piles for the sweepers to collect. Springfield Ranch, Beynon, Alberta. [ca. 1900]. Glenbow Archives, NC-43-43.

Figure 2.4. Haying in Pincher Creek area, Alberta, [ca.1900-1903] with an overshot stacker with two sweeps, which collect the mown and raked hay and haul it to the overshot stacker. Glenbow Archives, NA-2382-8.

market. The best females would be kept too but, on reaching maturity, bred and made part of the mother herd. The poorer ones would be neutered by spaying and eventually sold for beef with the steers.[6] Because the men so quickly fenced in their holdings they were able to realize important economies that had never been possible on the open range. As well as insulating their stock from inclement weather and protecting it from outside breeds, disease, wolves, and rustlers, they were able to time birthing so that their calves were born in the spring—March through April. In that period, the weather was likely to be moderate and birthing death losses small, and yet it was early enough to give the calves the full summer to grow and develop on their mother's milk and the prairie grasses. The calves were then weaned and converted to hay by the time winter set in.[7] Enclosed pastures also enabled the partners to make sure they missed none of the calves during the June roundups.[8] At the same time they were able to keep close enough watch on their animals to identify unproductive cows and sell them on the slaughter market when

FIGURE 2.5. The last stack at the Calf Camp. One man forks the hay to the other who places it and stomps it down to pack and finish off the top. *Rocking P Gazette*, October 1924, 20. Property of the Blades and Chattaway families and their descendants.

pulling off their "fats" during the September gatherings.[9] It was important to turn these animals into cash rather than allow them to absorb the precious grasses and the salt "licks" purchased to keep them healthy, as well as the hay they required when snow covered the ground.[10]

The partners kept a close watch on their cattle throughout this early period, grazing them most of the time and feeding them when necessary. By the summer of 1905 the herd had grown to 720 head. At that point, they made a monumental decision. They recognized that the area they had chosen was already occupied to a great extent, mainly by the Bar U ranch and the Bar S to their immediate south (then owned by Peter Muirhead), and that it was fast filling up with homesteaders. To expand in their current location would be a long, drawn-out process. They decided, therefore, to look elsewhere. To this end they took a huge leap of faith, and in 1905 they acquired a CPR lease on the Red Deer River that vaulted them from settlers to cattlemen in one fell swoop. It was too late in the season to do anything the first year, but by the summer of 1906 the future looked promising, and they stocked the lease with 1,500 head

The lease contained nearly two townships of land—37,342 acres; the cost was 40 cents per acre, for a total of $1,493.68 per year. In heading out to the Red Deer River area, they were following the lead of some well-known foothills ranchers, including their friend George Emerson, and also John Ware, both of whom had moved east to avoid the hordes of homesteaders pouring into the more fertile western regions; they were also following the example of future famous westerners George Lane of the Bar U and A. E. Cross of the A7, who leased new holdings to supplement their western pastures.[11]

At the same time, they took an opportunity to invest in freehold property in the region. The partners were looking for a place in the area on which to base their operation, when a tragedy occurred. While checking cattle on his range in September 1905, John Ware was thrown from his horse and killed. Ware had lost his wife, Mildred, to pneumonia some months earlier, and the Royal Trust Company of Calgary was appointed to take over negotiations to sell his property for his five orphaned children.[12] The partners were able to buy Ware's half section, N½ 19-21-14-W4, all his cattle, hay, and improvements, and his five brands, for $12,500.[13]

FIGURE 2.6. John and Mildred Ware and two of their children, Robert Lewis and Amanda Janet "Nettie," circa 1896. Glenbow Archives, 2-63-1.

Now Riddle and Macleay brothers had room to pasture one- to three-year-old steers that could be readied for market relatively quickly compared to newborn calves, which improved cash flows. They obtained most of these cattle from small farmers in Manitoba. Rod went east on one of many buying trips in 1906. The results can be traced along trails he left of cheques and receipts. A receipt written on 3 June on Albion Hotel paper in Portage la Prairie reads: "From C Knox, 627 head, 3 year olds

$25.50, yearlings $19.50." The freighting charge was $40.00 per car from Winnipeg to Brooks. Cost per head landed at Brooks was $21.29 plus $1 commission. Charlie Knox was a well-known cattle buyer in the West and one of the few who competed for the cattlemen's business with the West's wealthiest cattle buyer and slaughter plant owner, Patrick Burns of Calgary, and with the beef brokering firm that regularly colluded with him, Gordon, Ironside and Fares out of Winnipeg.[14]

Initially Uncle John Riddle helped finance some of the cattle. A letter arrived in May 1906 advising Rod that there would be a credit of $6,000 at the Bank of Montreal for the purpose. It suggested purchasing yearlings, as they would be able to buy a greater number than if they bought older, heavier stock. With this money and advice Rod went back to Winnipeg and got five carloads of "doggies."[15] These cattle cost $22 per head, plus the one-dollar commission to Knox. On 1 June 1906, Uncle John brought in 300 steers of his own and had them branded with Bar M7 on the right rib and then turned out on the Brooks lease. He was helping the partners financially by subleasing grazing privileges from them.

Since all the steers came from Manitoba rather than Ontario, where many western ranchers got their supply in those days, the partners' herd would have been of an acceptable if not the highest quality.[16] Manitoba milk producers normally bred their cows to Shorthorn, Hereford, and Angus bulls rather than the lankier dairy varieties, including Holsteins and Ayrshires, that Ontario farmers often used. Consequently, the offspring were only 50 percent dairy bred and "fleshed up" and put on weight comparatively more efficiently. It was likely Emerson's tutelage that accounts for this decision. In 1900, a neighbouring rancher had remarked: "G[eorge] Emerson of High River bought over one thousand [Manitoba cattle] this year. He has been buying for some years and seems satisfied with results."[17]

The beef market in these years was on a very slow but somewhat steady climb, which suggests Riddle and Macleay brothers probably made, or at least did not lose, money buying doggies and pasturing and selling them within one to three years as "fats." The average price of slaughter steers nationwide between 1901 and 1905 rose from $.0436 to $.0452 per pound.[18] The one-year-olds landed at the Ware ranch at $27.00 per head (transportation and commission included) would probably have weighed 400 to 500 pounds. If the grass was lush and the partners grazed the animals for 100 days that summer, wintered them, and then grazed them for 100 days again the next year, they would have been able to put around 600 pounds on each of them over two and a half years.[19] If they sold them at the average price of about $.0452/lb. they would have got 1300 × $.0452 or $58.76 per head on the fall market. Of course, there were costs such as for death losses, which might be around 5 percent, for the CPR transporting the stock to market, for hay over the winter, and for incidentals like salt. Much of the labour was their own, but more and more they were relying on employees, especially on the Red Deer River property. By 1907, besides the three partners, there were up to twelve men on the payroll. This clearly separates the partnership from the average family farm on which virtually all labour was home-grown. However, as we will see, family strategies and practices would also be a major part of their operation's future.

To say as much is not to argue that Riddle and Macleay brothers had money to spare. Overall startup costs were high. They had invested in cattle, lumber, and provisions before coming west and then had built the home place, purchased barbed wire for their fences, and paid for all the necessities of life. For the time being, they had to maintain their breeding stock while they waited for the calves born to their own cows to mature, fatten up properly, and eventually contribute to their liquid position; and, of course, they not only required capital to bring in doggies but also to make payments on land purchased and leases. This clearly put them in the realm of borrowers far beyond the average settler. As a consequence, the partners felt obliged to diversify their forms of production. In so doing, as we will now see, they were to be setting (and following) what was to become a well-trodden path for their settler kind in western prairie Canada.

3

Nature's Fury and the Tattered Dream

Riddle and Macleay brothers were not to last much longer as a ranching partnership, but until the infamous winter of 1906/7, they seem to have maintained their enthusiasm. This they needed in the search for creative means to improve their cash position.[1] Like many of the settler ranchers, the first means they undertook was to diversify their forms of production. That endeavour soon brought them into the horse business. Horses were usually worth more per head than cattle, and they wintered more easily, in part because, unlike the latter, they had the good sense to "paw" through the layers of snow to get at the pickings below.[2] Ranchers in the Canadian West had access to good working ponies from the beginning that came north from Montana with the first cattle drives. Many of these horses had descended from Texan stock. They were the ones often referred to as "cayuses"—a mixed breed that was more a product of function than design, had survived on the open range since the eighteenth century, and away from which the weaker specimens had been pruned by Mother Nature. As the first cattle frontier had moved north along the eastern edge of the Rocky Mountains, ranchers had bred these relatively small (and fast) steeds with larger varieties, notably Thoroughbreds, Irish Hunters, and draft breeds that had been brought in from the East and overseas via the transcontinental railways. The progeny were relatively large and could carry a cowboy loaded with heavy winter clothing and camping gear through deep coverings of snow. They were sturdier than the cayuses. Though they would not have shown very well against prized animals in eastern auction or show rings, they were exactly what was needed to open the West. The heavier set of these animals were also good for draft as well as riding purposes, and when mixed and grain farmers began settling the northern extremities of the Great Plains they relied on them to plough up the virgin prairie soils.[3]

It was the wheat boom in the early years of the new century that really augmented this demand. Prior to 1905, wheat averaged around 80 cents a bushel on the Winnipeg Grain Exchange; then from 1905 to 1914 it spiralled upward to over $1.30, and through the Great War to $2.24.[4] To numerous cash-strapped settlers responding to the federal government's aggressive western settlement campaign, grain farming was a means of exacting a quick return without investing in fences or barns or haying equipment or waiting three to four years for newborn calves or even one to three years for eastern doggies to grow and fatten. Grain yields on the flatter plains around High River to the north of the Porcupine Hills grew from zero bushels in 1904 to 99,800 in 1905 and then to a massive 600,000 bushels in 1906.[5] In a short time, High River, just twenty miles from Riddle and Macleay brothers' home place, changed from the centre of cattle ranching in southern Alberta to the largest individual grain shipment point in western Canada.[6] Seeing the tide of migrating farmers and resisting the dubious complaint of many cattlemen about being crowded off the range, Rod, his brother, and cousin decided to sell horsepower whenever they could. To get established in the business, Rod made his first trip south in the fall of 1902 to the more mature and established ranching country in Oregon, primarily to purchase brood mares.[7]

Riddle and Macleay brothers pastured the mares in the summertime with well-bred stallions and raised the foals in fenced pastures summer and winter, where they could protect them and even occasionally supply them with feed when the weather was severe and snow particularly deep. The best fillies they kept as replacements for the mares, and the rest, along with the geldings, they sold. They also gave the animals the individual attention needed to prepare them for the market. Each year they spent much of the latter part of June acquainting the horses with the finer points of riding and hauling. The men treated every outing as a means to this end. Even for a fishing trip to the south fork of the Highwood River (Pekisko Creek) to the north, or Willow Creek to the south, or a

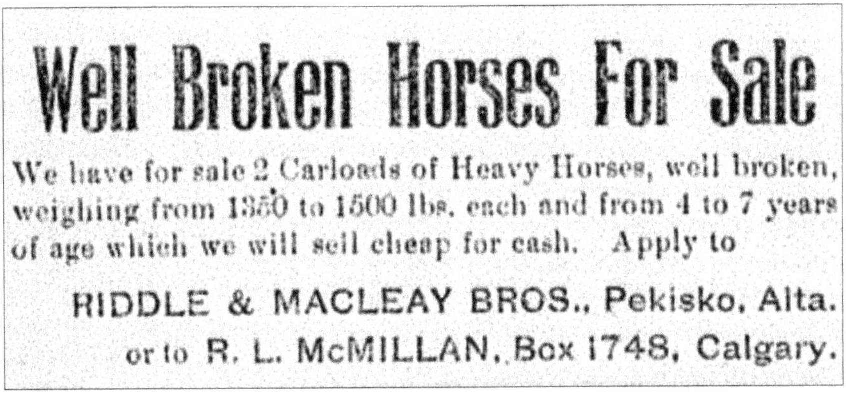

Well Broken Horses For Sale

We have for sale 2 Carloads of Heavy Horses, well broken, weighing from 1350 to 1500 lbs. each and from 4 to 7 years of age which we will sell cheap for cash. Apply to

RIDDLE & MACLEAY BROS., Pekisko, Alta.

or to **R. L. McMILLAN, Box 1748, Calgary.**

FIGURE 3.1. Advertising the horses. *Gleichen Call*, 29 August 1907. See "Peel's Prairie Provinces," University of Alberta Libraries, Page 3, Item Ad00302_1.

quick jaunt to pick up mail at the Bar U, they saddled up or harnessed green broke horses.

In 1907, Rod Macleay flipped ninety head of Bar N horses he bought from an American commission firm, Parslow & Hamilton, for a profit of over $11,000. At this point, all across the United States, cities were moving away from horse-drawn public conveyance and there was a glut of draft horses. Rod's growing familiarity with markets, commission firms, and railway shipping enabled them to take advantage of the situation by purchasing broke, heavy horses to sell to the growing farm population desperate for horsepower. For six weeks, 22 August to 26 September 1907, Riddle & Macleay advertised the stock in the local newspapers.

While marketing the American horses, Rod also went out to Arrowwood, Alberta, south of the town of Gleichen, and bought thirty-seven colts from Chris Bartch for $4,275. In April, he bought a team from Billy Henry for $250. However, to acquire enough supply he had to continue to do some of his dealing across the line. A buying expedition he made back to Oregon in March 1910 shows the complexities of shipping stock by rail in those early days. In the family files are inspection certificates dated 8 March 1910 signed by veterinary surgeon J. A. Donaghue, the local state stock inspector at Baker City. He examined ninety-one horses for "Roderick Macleay of Cayley, Alberta" and certified them "free from contagious and infectious diseases." Even then bureaucratic procedures had to be followed. Transportation costs were high, both monetarily and in terms of wear and tear on the animals. The horses rode the Sumpter Valley Railway Company train from Baker City, Oregon, to Spokane, where they were transferred to the Oregon Railroad and Navigation Company Railway. From Spokane they rode to Shelby, Montana, and then via the Great Northern to Sweet Grass, where they were transferred to the Alberta Railway and Irrigation Line to Lethbridge and finally the Canadian Pacific Railway to Cayley. The duty was $837.50 and the total cost for purchasing and shipping 101 horses was $9,117.75, or $90.27 per head.[8]

Macleay's notes show that this was a very hard trip for the stock: "Left Baker City, 3 cars went off track at Telocaset, Oregon; delayed 4 hours. Thirty-five miles further on 3 more cars in the ditch at Meacham. They were Armour refrigerator cars loaded with Armour's very best canned ham, tongue, Irish stew and pigs feet; we all had a good lunch.[9] Delayed 4½ hours. Walla Walla is a hell of a hole." Later, Rod brought a suit against the Oregon Railroad and Navigation Company and the Northern Pacific Railroad for failing to feed and water the horses between Baker City and Sweet Grass. In the end the horses recovered with no lasting effects, but Rod was determined to make a point. Had the trip been made in the heat of the summer months there could have been severe losses.

Despite competition starting as early as the first decade of the twentieth century from steam and then gasoline tractors, draft or work horses were to be important in western grain and mixed farming until after World War II; and, of course, even today, good equine stock is utilized by ranchers and any rancher/farmers who incorporate a significant grazing component into their program. Rod Macleay would thus continue in the business well into the 1940s. The draft horse eventually became obsolete, but it is interesting that many farmers were still making at least limited use of them in the decade or so after the war.

Like many other frontier ranchers and farmers, Rod Macleay instinctively recognized the need to work closely with his neighbours. He apparently had accepted an obligation to supply a rider to the Bar U roundup on what must have been some still unclaimed communal range. On 26 October 1904, he took his bedroll and his horses Stub, Noah,

Brownie, and Dick, and he borrowed three more from Herb Miller, the foreman of the big ranch. The Bar U normally hired about fifteen riders to help round up its stock, with 7U Brown as "wagon boss."[10] Between 28 October and 17 November, the cowboys gathered some 7,000 head. They would move to a new area of the rangeland each day and then ride out some twelve miles before sunup and sweep back, gathering all the cattle along the way. When Rod was finally turned loose from the job, he was very glad to get back to the home place. It was cold and the cowboys' tents had been blown down quite regularly by the high winds.[11]

One year, after settling their own cattle for the oncoming cold season, all three partners headed out to the interior of British Columbia with fellow rancher R. L. McMillan, who "took on the job of secretary and bookkeeper," to do some logging for the Canadian Pacific Railway. Rod's younger daughter, Maxine, born in 1911, explains: "to obtain some income the boys moved workhorses out to the Cranbrook area, obtained a contract, and spent the winter getting out railroad ties." They acquired "a timber berth" of their own "near St. Leon's south of Revelstoke on Arrow Lake,[12] Logging was a familiar undertaking to the three men, as they had all been exposed to the timber industry back in Quebec. Rod and Alex's father had run the St. Remi Lumber Company in the Eastern Townships, and Rod himself had worked at the mill as a bookkeeper.

Unfortunately, the logging undertaking turned out to be a bust and they barely broke even. In the summer of 1906, however, the future looked promising. Riddle and Macleay brothers could take some satisfaction in the knowledge that they had made a substantial investment in both land and cattle, and, as Rod's diary demonstrates, he had a vision of land prices greatly increasing and of a strong horse market as the settlement process continued. Unfortunately, Mother Nature was about to make that possibility seem much less certain. The fortunes of cattle grazing depend to a significant degree on the vicissitudes of weather. The two major climatic problems one has to face from time to time in southern Alberta are drought and vicious winter storms. The home place was in a part of the foothills region where moisture is more abundant and drought is less severe than in other areas. Out on the Red Deer River it can be more severe, but cattlemen are able to adjust their numbers there to offset any pasture depletion resulting from it—usually by bringing some or all of their stock home earlier than usual in the fall. Winter weather, on the other hand, can be devastating in both regions. In the fall of 1906 through to the spring of 1907 it turned viciously against everyone grazing stock in the Canadian West and the American Northwest. Canadian historians have downplayed the impact of this celebrated winter, but virtually every eyewitness report one reads emphasizes both the devastation it inflicted and its breadth.[13]

It all started on 15 November when rain that had been falling for two weeks suddenly turned to snow and the temperature plummeted to 15 degrees below zero Fahrenheit. Some three feet of snow fell in a few hours. Then the temperature climbed above freezing for a few hours and quickly dropped again, forming a layer of hard crust under the fresh snow that made it even more difficult for the cattle to graze. One blizzard followed another until late spring. A number of the big cattle owners were still grazing stock year-round on the open range. This winter proved the end for many. Their animals soon began to die from starvation and cold. Many cattle in the foothills pushed south and east in a futile attempt to escape the northern winds and to search for food. This left them on the open plains without the wind protection they could have got from the cutbacks and forests in the high country. Some ranchers rode out into the storms to attempt to hold the cattle back, but in vain as the animals flowed around and past them like a mighty river. Other ranchers tried gathering them in bunches out on the plains to drive them back up into the hills. One such rancher recalled, "Think of riding all day in a blinding snowstorm, the temperature fifty and sixty below zero, and no dinner. You'd get one bunch of cattle up the hill, and another one would be coming down behind you, and it was all so slow, plunging after them through the deep snow that way; you'd have to fight every step of the road." The horses' lower legs "were cut and bleeding from the heavy crust, and the cattle had the hair and hide wore off their legs to the knees and hocks. It was surely hell to see big four-year-old steers just able to stagger along."[14]

Finally, the exhausted riders and horses had to just let the cattle go. That sealed their fate. "Fence corners, railway tracks, coulees, river bottoms" filled up with the bodies of dead cattle. "One day in January the citizens of Macleod saw what appeared to be a low, black cloud above the

FIGURE 3.2. Dead cattle, Shaddock Boys Ranch, Langdon, Alberta, Spring 1907. Glenbow Archives, NA-1636-1.

snow to the north, which drew slowly, draggingly [sic] nearer until it was seen that a herd of thousands of suffering range cattle were coming from the north, staggering blindly along the road allowances in search of open places in which to feed." Painfully "into the town this horde of perishing brutes slowly crawled, travelling six and eight abreast, bellowing and lowing weak, awful appeals which no one was able to satisfy." Those that made it through the town surged "out into the blackness of the prairie beyond, where they were swallowed up and never heard of again, every head being doubtless dead before the [next] week had passed."[15] Before they died the poor brutes ate everything in their path—small sapling trees sticking through the snow, the hair off the backs of one another.[16] In the Milk River area near the United States border, "there were so many dead cattle" dotting the landscape the next spring that one young lady who was still relatively unfamiliar with the countryside "found them very useful" as landmarks for making her "way about the prairies."[17]

Something like 50 percent of the cattle on the ranges froze or starved to death during the endless march of blizzards.[18]

The cattle people who survived this dreadful winter were the settlers who had fenced off their ranges and could keep their cattle close to forests and other natural protection and to sufficient amounts of the precious feed they had stockpiled the previous summer. In 1907, the commanding officer of the Royal North West Mounted Police at Macleod reported: "Last winter was an exceptionally long and cold one. It was said to be the coldest in twenty years. Cattle in consequence suffered a great deal, and large losses had to be recorded, especially by owners of large herds who could not feed and look after their stock the way the small owners could. These latter suffered very insignificant losses."[19]

Rod Macleay and partners were now "small owners" only in comparison to the company outfits. Like other family operations, however, they had, as we have seen, fenced a good deal of their pasture land and put up substantial quantities of hay. Consequently, they were able to nurture most of the cattle they had grazing at and around the home place. Their losses in that region were not large. On the other hand, having not yet experienced a really severe winter, they took a chance on the open range in the Red Deer River area. There they lost heavily. We do have detailed information directly relative to cattle A. E. Cross had running in that area that is contained in the correspondence that he kept up from Calgary during the course of the winter with his hired man, Charlie Douglass. It probably provides a very close approximation of the fate the Riddle and Macleay brothers' stock suffered.

Once the storms began, Douglass and Cross's other hired man, Billy Maclean, spent much of their time in temperatures of 30 and 40 below zero Fahrenheit. They first fed up their supplies of hay, and then they hauled hay, chopped oats, and greenfeed to the cattle that they were able to purchase from neighbours. On a particular morning in November they were both almost killed when they were caught in a sudden storm. That morning Maclean went out alone to do some feeding. About midway through the morning it started to snow, and the temperature, which for a short time had been relatively mild, suddenly plunged. He was probably unprepared and poorly dressed and, suffering from hypothermia, he left his horses and wagon at the stacks and started out across the prairie on foot. At noon, when he failed to make it back to the house, Douglass went looking for him. He soon came across Maclean, pulled him up on his horse and headed for home. The blizzard was fierce and Douglass found that he could not head into the mind-numbing wind. He made his way instead to the Red Deer River and then followed it under the protection of the steep valley side until he reached the house. They were two lucky men. Within three hours one side of the house was completely covered by a gigantic snow drift.[20]

When this storm let up a bit, the men realized they were fighting an enormous battle.[21] On 20 January Douglass wrote that the losses were bound to be heavy as there were no fences and the cattle had drifted off their respective pastures into the river valley and were all mixed up. Those ranchers who "have hay can't get their cattle to it."[22] Douglass had two teams hauling feed to about 300 head at this time, and he thought he had enough to last until the end of January. A letter of 27 January demonstrates that this situation quickly deteriorated. The cold weather and blowing snow made it increasingly difficult to haul feed. The cattle along the river were starving and eating brush. Douglass now had about 100 head of Cross's cattle in the field and "this is all I can possibly manage so it's no use looking for the poorest any more."[23] He reported on 16 March that they were skinning the hides off the dead animals and selling them for six cents a pound. He realized that this was not very profitable, but he offered the commiserating comment that at least it would help to defray the cost of some of the chop they had been using for feed.[24]

Cross's herd along the Red Deer River was decimated. He had started the winter with eight hundred cattle on his lease and by spring he had "two hundred and fifty head left."[25] For Riddle and Macleay brothers the losses were larger in that they had almost double the number, just under 1,500 head—all dry stock from the home place and 100 purchased, totalling 500-odd head, 300 Manitoba doggies of Uncle John's, and 627 of mixed-age cattle from Manitoba for the partnership. The largest percentage were yearlings. In later years Rod was to hire a neighbour named Sam Howe to foreman a crew to watch over the cattle year-round.[26] At this time, however, the partners thought they could run that part of their operation with the model the great corporations used. Months after the storms abated, Pete Muirhead, who ranched near their home place in the Porcupine Hills, wrote:

> We had a very bad spring here. The worst we have had yet after a hard winter. The loss in Cattle was large. The Round-up wagons have returned after gathering cattle on the range. They were as far south as the Montana line and the loss was larger than they expected. Riddle & Macleay Bros. who live three and one half miles north of me had a lease on the Red Deer River and had 1500 cattle on the lease. Two hundred and fifty were all they had left. And out of eight hundred calves that they had, they are all gone.[27]

Over the next few years, beef prices plummeted in the West as ranchers throughout the northern Great Plains threw what remained of their stock on the market and quit. Rod's partners too were discouraged. Things had changed a great deal in these first eight years. The greatest change, over and above their financial setbacks, was Rod's marriage. He had met and married Laura Sturtevant while on a short visit to Burlington, Vermont, in late 1905 to celebrate the engagement of his older brother, Dr. Kenneth Macleay, to Laura's sister Gertrude. He brought Laura out west in January 1906 to live in the house he and the others had built, and about the time his partners decided to quit, the couple were expecting the first of their two children.[28] This would make the house even more crowded and, quite naturally, Alex and Douglas wanted, and were actively pursuing, wives of their own. For them, the writing was on the wall: they had outgrown the partnership and they soon let Rod know they wanted to get out. Fortunately for them, Macleay decided, no doubt after considerable soul searching, that he wanted to keep going.

Rod knew he was not in a position financially to buy his partners out. He needed a backer and, inevitably, he turned to his mentor, George Emerson. Years later daughter number two, Maxine, described what happened. "When Dad wanted to buy out Douglas and Alex, he rode from the 'Nigger John'[29] as the ranch was called, to Emerson's ranch [now on the Bow River near Redcliff], and asked to borrow money ... George said: 'No, I won't lend you the money Rod, but I'll buy them out

FIGURE 3.3. 1948, L-R: George Chattaway, Roderick Macleay, Stewart Riddle, Charlie Glass. Last bunch of Rocking P beef steers trailed in to Cayley. After this date, steers were taken by trucks to the Calgary stockyards. Glass "was employed by Macleay in the early 1920s ... and stayed for many years" eventually becoming foreman of the Rocking P (Henry C. Klassen, "A Century of Ranching at the Rocking P and Bar S.," *Cowboys, Ranchers and the Cattle Business: Cross-Border Perspectives on Ranching History*, ed. Simon Evans, Sarah Carter, and Bill Yeo. [Calgary: University of Calgary Press, 2000], 112). He does not seem to have been working at the Rocking P during the months when the Gazette was produced. Glenbow Archives, NA-3535-191.

myself and we'll be partners.' So Dad hitched up the team and leading his horse [behind the wagon,] the two drove to High River and made a deal." Macleay's diary entry for that day reads: "Bought out Riddle & Macleay Bros. lock, stock and barrel." Macleay was named working manager, and partnership cattle were to be branded with Emerson's Rocking P brand.[30] It was from this point on that the ranch at the home place increasingly assumed the appellation the "Rocking P."

The Riddle and Macleay brothers extended family period thus came to a relatively abrupt end. Considering the fact that it ended on a sour note financially speaking, it could be seen as a period of abject failure. To judge it as such, however, would be somewhat misleading. Rod's share of an expanded partnership had grown to 50 percent, and he and Laura were the sole proprietors of the home place. But perhaps most importantly, it was out of this early stage that Rod himself carried with him some important lessons that would help him survive in the ranching business, at times against rather formidable odds. The most important of these lessons were respect for Mother Nature and the need to be flexible, that is, to adapt to both market and financial circumstances as they presented themselves.

4

The Rocking P Ranch (and Farm)

Macleay and Emerson wasted little time in rebuilding their cattle inventory. They may have made out reasonably well financially after 1909 as the price of cattle on the depleted southern Alberta (and Saskatchewan) pastures rebounded.[1] However, their partnership lasted just six years. This was largely because of Emerson's age (he was born in 1841); but according to local historian Lillian Knupp, it was also because the two men's business philosophies were very different.[2] Macleay was a "hard-nosed" businessman who was willing to pursue any agricultural practices that were likely to make money, while Emerson still embodied in many real ways the transition of the region from a fur trade and open bison range into one characterized by free-roaming herds of cattle and horses. Emerson was informal in his approach, to say the least. He managed largely by an "uncanny sense" and a "keen eye" but either spurned or resisted the more careful business techniques. He "kept no records of his ... dealings." After his partnership with Macleay ended he apparently "took all of Rod's carefully kept accounts for the [past] six years ... opened the lid of the stove and stuffed them in; the deal was closed." The Gordon, Ironside and Fares beef conglomerate out of Winnipeg once made inquiries about a $5,000 cheque it had issued to Emerson three years previously for cattle the firm had bought and shipped to Chicago. George "searched his belongings [and] found it wadded up in his vest pocket, all worn and tobacco stained."[3]

Knupp's descriptions say as much about Rod Macleay as George Emerson. Macleay was meticulous in all his dealings; and this would eventually prove instrumental in the development, expansion, and, indeed, the survival, of his entire operation. He was also a fast learner. Working from 1914 on without outside partners for the first time, he never again left unattended cattle on the Red Deer range. He shipped any cattle from his eastern pastures that were ready for the market before winter set in and trailed the younger and more delicate stock back to the home place where it could be watched and nurtured closely.[4] Eventually, he left some cattle, presumably more mature and reasonably hardy two- and three-year-old steers, on the Red Deer, and he hired someone to see to the care of the stock. Also, as farming increased in the area so did the availability of feed in the form of the straw piles the farmers left in the fields after harvest. In those days threshing machines were stationary when in operation and grossly inefficient compared to modern combines. The excess straw the machines cast off as waste always had more kernels of shelled grain left in it than would be considered acceptable today. When a farmer and rancher could work out a realistic agreement to utilize this otherwise worthless residue as a form of winter grazing rather than leave it to rot in the fields, it just made common sense. A foreman and crew would spend the winter moving the cattle from pile to pile. This system had the effect of expanding Macleay's land base without requiring him to invest in the costs of ownership. At the home place he fed his cattle on his own hay, but he also paid local farmers to pasture some of the stock on their straw piles. Farmers, both there and on the eastern ranges, actually even assumed responsibility for cutting water holes in ice-covered sloughs, lakes, or streams for the cattle and for moving them as necessary to the natural shelter of a valley or patch of trees.[5]

Macleay's inclination to work out these arrangements with his neighbours indicates his practical approach to the grazing industry. He helped them market their excess supplies of grain, straw, and hay, and they in turn saw to the welfare of his stock during the most difficult time of the year. He was similarly pragmatic with regard to the forms of agricultural production he was prepared to embrace. When it made business sense, he was willing to diversify beyond cattle, and now the horse business, into any sub-industry he could think of. In other words, he was willing to operate like so many other family outfits that sustained their livelihood through the first half of the twentieth century in western prairie

Canada. The *Rocking P Gazette* speaks of chickens and eggs on the ranch as well as milk cows.⁶ These were kept to supply the country table. Hogs, also mentioned in the *Gazette*, contributed to the table too but, more importantly, were a means of making use of substandard grain.⁷

Rod also did not balk at tilling the soil and harvesting grain in the search for cash to finance his debts. After purchasing the Bar S ranch in 1919, which had a sizeable farming component, he committed to grain production.⁸ He seeded down some of the more arable land around the home place. Two of the regular workers, "Clem and Val, left here Nov. 13th … to help thresh," the *Rocking P Gazette* reported in November 1923, and "after a strenuous week of work they returned on Nov. 19th, with two loads of oats, which filled the out-bin, and also the bunkhouse."⁹ Later the same month one of the men "celebrated the finish of the threshing, by staying in bed until 9 a.m.," the *Gazette* jokingly noted. "When he arose he" luxuriated "by taking a bath." He completed "his toilet by using 'Florida Water,' and massaging his face, in some oriental solution, with much admired results."¹⁰

The oats were used principally to feed Rocking P horses. However, Macleay also planted and harvested wheat as a cash crop to bolster his beef sales. During the post–World War I depression from 1920 through 1925, beef prices declined dramatically. Wheat fell on the world market too, but it bottomed out at about a dollar a bushel and oscillated upward to as much as a dollar and a half. In that range, it was more or less profitable. Two of the ranch hands, "T[ommy] McKinnon and Jimmy Hendrie, returned two grain-tanks [filled with wheat] to the High River Wheat & Cattle Co. on Oct 5th," the *Gazette* reported in 1924.¹¹ Grain tanks carried twice the load of the average wagon. Producers used them to deliver their grain to a shipping point where an elevator company then sent it by rail to the East and overseas by steamer. On 25 October 1924, hired man Ed Orvis "brought two new grain tanks out from Cayley" to be filled and returned.¹²

Macleay and others like him realized what ranchers before them never understood. Ranch sustainability depends in no small way on the preservation of the natural environment. The way to get the best production out of the land was to treat it with respect. However, the concept of sustainability was not applied until the drought of the 1930s forced

FIGURE 4.1. A cartoon sketch by Maxine Macleay, *Rocking P Gazette*, October 1924, 19. Property of the Blades and Chattaway families and their descendants.

the issue. Grassland management and sustainable farming techniques are still an evolving science, but at the time there was a simple recognition that when something was removed it should be put back. In the 1930s, Rod was using phosphate fertilizer on the 4,000 acres of land he was cropping annually. Lately, artificial fertilizers have got a poor press from environmentalists, principally because of the residue they leave particularly in our natural water sources.[13] In the 1930s, however, the latter problem was not understood and fertilizers were rightly considered a way to restore much of the nutritional ingredients field crops were removing. Macleay "carefully studies agriculture (not just cattle)," the *Lethbridge Herald* reported, and in 1931, "after 3 years of study … he used a carload of phosphate fertilizer on his grain land." This "increased the yield to 10 bushels to the acre" and ripened the grain "10 days earlier." By then grain was contributing directly to beef production. Macleay was finding barley more suitable to the short growing season in the hills than wheat, and according to the *Herald*, he was using virtually all of it to grain finish cattle in his "feedlots."[14] "All of the many corrals … are supplied with running water," the paper added. "Each … is well-staffed; most of the 'hands' having been handling [Macleay stock] for years."

The above newspaper article reported that retailers in Montreal were featuring Macleay's well-marbled beef in their ads.[15] Over the years, Macleay also attempted to market some cattle internationally. He first exported to the United States in 1907 and then to the United Kingdom in 1910. As the Great Depression set in the late 1920s and early 1930s he expended considerable effort to get the Western Stock Growers' Association, the Canadian Council of Beef Producers and the department of Agriculture in Ottawa to lend their support to the British trade.[16] He was successful in that endeavor and from 1930 through 1933 he also shipped more of his own cattle to that market. He knew very well what the buyers wanted and, along with other reputable ranchers like the McIntyres and A. E. Cross, who controlled their breeding and had enough cattle to be selective, often impressed British buyers. Evidence from Macleay's sales provides yet more examples of the advantages the new family ranchers had over the former corporations that had dominated the first cattle frontier. It was a new era with new rules. In the earliest years, the United Kingdom had been the target market for surplus cattle, and only grass-fattened animals were sent from the Canadian west to the markets there. Grass fat tends to be "soft" and to "shrink" away rather badly when the animals are pulled off the ranges. The corporation cattle had thus tended to lose much of their weight on the long trips by rail and steamer to the United Kingdom. One expert writing in the 1890s put it as follows: "cattle wild, excitable and soft off grass, are driven to the railway, held sometimes for days on poor pasture waiting for cars, and finally, after more or less unavoidably rough handling, are forced on board" an ocean steamer. "After a journey of five thousand miles ... our grass-fed range steers arrive in British lairages [sic] gaunt and shrunken, looking more like stockers than beeves," and the British "think we have no feed."[17]

The other problem for the company ranches was the impossibility of achieving efficient breed selection at a time when on the open range any bulls could access any cows at any time of the year. Scrub bulls of low quality had roamed widely and constantly competed with any better beef bulls the ranchers introduced. Even after the corporate ranching era, the problem of poor-quality cattle continued. Part of the problem was that the producers who made an effort to raise a better quality were not properly recompensed, in part because of the lack of a modern grading system. Consequently, the quality of progeny, whether properly finished or not, was always relatively low compared to the best British animals. Time and again in the 1880s and 1890s reporters attending the auctions at Liverpool, Manchester, and Glasgow had commented: the Canadian cattle "were of a middling and ordinary quality;[18] "from abroad the supplies of stock consisted of 700 cattle from Canada which were a moderate lot. Some of these were taken for keep [i.e. feeding], the rougher description meeting the worst trade of the season, entailing heavy losses for the exporters;"[19]

In entering the British markets, Macleay attempted to overcome both these problems. Obviously, he had the means to produce well-finished beef. Moreover, as the photographs below help to demonstrate, by this time he had been able to raise the quality of his stock to the highest possible level, mainly by working year after year to upgrade his breeding program. His grandson offers the following elaborate description of his approach:[20]

Rod had no preference for breed, or colour, but he did recognize hybrid vigor. He was not biased, one way or the other, and never expounded on the merits of any single breed. He practiced cross breeding, whether by design or by accident, before anyone else, and as usual, it was contrary to the popular trend. He generally had a mixed battery of bulls consisting of just about all the beef breeds available. Basically, by 1930, the herd looked Hereford but there were Angus, Galloway, Shorthorn and even two imported Highland bulls. The number per breed varied from year to year. [His] ... breeding plan ... can be best described as mixed breeding, but it was still cross breeding and it was not very fashionable. He maintained this approach throughout his career when the popular consensus was to "straight" breed, with any breed, and the closer you got to purebred the better. Anything showing mixed blood was a mongrel.

His initial herd was Shorthorn, like everyone else's, because there was little choice if you needed numbers. After ... WWI, he used a lot of Hereford bulls, which by that time were the most common breed available. He was partial to roan cattle and in 1928 he decided to swing back to Shorthorn, in a big way of course, but it would come at an enormous ... cost for one year. He bought one Hereford from Walter Davis and one Her[e]ford from Charlie Lehr but the vast majority were Shorthorns. A whopping 28 came from the Calgary bull sale at a cost of $280 a head and then 19 more came from Mr. Dryden of Brooklyn, Ontario for $175 a head. This was a huge replacement rate nearly 50 head, all the same age and amounting to nearly 50 per cent of the bull battery. He did a similar thing back in 1917–18, when he went to Her[e]ford buying 54 head within a year. A normal annual replacement rate would be about 20 per cent annually. The 1928 purchases would all have to be replaced about the same time and he would have to do it all over again. He must have regretted not doing it two years earlier because it could have been done at half the cost.[21]

> # Alberta Varsity Shows World A Few Things About Cattle
>
> Department of Animal Husbandry is Developing Shorthorn, Hereford and Aberdeen-Angus Breeds Which Take First Place in the Markets of North America

FIGURE 4.2. Recognizing theat Shorthorn, Hereford and Angus were numerically by far the dominant breeds, not just in Albertan but North American beef cattle operations generally, the Husbandry Department at the University of Alberta saw fit in 1928 to specialize in the scientific development of those three breeds alone. *The Gateway*, 8 November, 1928, 1. See "Peel's Prairie Provinces," University of Alberta Libraries, Page 1, Item Ar00103.

The *Rocking P Gazette* reported in April 1925 that "R Macleay bought eleven head of bulls at the Calgary Bull Sale, and they arrived home on the 12th the cowpunchers being Val Blake and Ted Nelson."[22] By mixing top bulls with his best cows Macleay was able to produce better offspring and improve his herd overall. As noted above, to ensure that poorer heifer calves did not hurt herd quality he subjected them to the spaying "hook" before fattening them out.[23]

The Old-World dealers and auctioneers were most impressed with the Macleay cattle. "Better by a long way," than many others being offered, one of the latter noted on 29 November 1930. "Some sold this a.m. for 20 pounds a head, some 19 pounds, 8 shillings and 5 pence" which was pretty much the top of the price range at that point.[24] The British buyers were used to well-bred Hereford, Angus, and Shorthorn stock in their sales rings—the type that most consistently provided highly finished beef—and they would not have been impressed with anything less.[25] The commentator was clearly lecturing Canadian producers when he noted that evidently all they had to do was "send a good beast and he will get a market." His animals will compete favourably "with Irish & English cattle."

FIGURE 4.3. Spaying heifer on scaffold, Maple Creek area, Saskatchewan, 1897. Glenbow Archives, NA-3811-96.

The Canadian government and, presumably, a lot of the other western beef producers, were thankful to those who diverted trade to Britain because it took some of the pressure off the North American market and for a while at least seemed to support prices. Following is a transcription in the Macleay family papers of a letter from the Department of Agriculture to Rod Macleay.

> On the whole, Dept is of the opinion that exporting of these cattle has beneficially affected the Western Market situation. … Stiffening of prices on market actually materialized when

shipments were in progress. Those who consigned cattle in connection with these shipments will have contributed greatly to betterment of market situation. We appreciate the part you have played in this effort to help the industry and hope it may be as much advantage to you personally as it apparently has been to the average beef producer in this country.[26]

However, Macleay's characteristically meticulous records of costs and returns told him that this lot of cattle lost between $.61 and $1.41 a head in comparison to what he felt he could have got for them in Canada. There can be little doubt that part of the difference was the excess shrinkage to which grass-fattened cattle were subject. Following is a set of Macleay's records for this trip.

Nov 1, 1930

Steers weight 1220 lbs per head when loaded at Brooks
Value at Brooks @ 5 cents per pound was $61.00/head
Net returns of Manchester was $60.39/head
Loss per head   0.61/head

CPR freight to Montreal $900.88
Stock Yards Montreal 138.17
Ocean feed and bedding 145.72
Ocean freight @15.00 each   870.00
Misc etc – total   2190.28
Comm etc Manchester 194.96

Total Charges $2385.2.[27]

From 1931 through to 1933, therefore, Macleay diversified his approach. He sold some grass-fattened cattle domestically and shipped two other types overseas. The first were "store" (or feeder) steers off the grass that were young—mostly two-year-olds—and still lean or (as the cattlemen would say) "green" enough that they did not have a lot of flesh on their

FIGURE 4.4. Preparing to load cattle on rail cars at Cayley stockyards. The white faces reflect Hereford and possibly Hereford/Angus crosses, which British buyers desired. These cattle were definitely not well finished and would, therefore, be offered for sale as feeders. Photograph property of the Blades and Chattaway families and their descendants.

carcasses to lose during the long journey. They targeted the feeder market. British farmers liked these cattle too, because once settled in they fattened up rapidly on their lush grasses, abundant supplies of corn, and turnips. The other type Macleay shipped were three-year-old steers that he had fattened on his own barley and feed wheat. He sent 307 head on 13 May 1932 and another 40 on 5 July 1932. These cattle were all cut out of a set of 585 he "had fed through" the previous winter. They were "a good lot" and "well finished" by the standards of the day. They weighed out very well considering the stresses. The one set we know of netted 1228 pounds, which was excellent, and brought $.05 per pound, or $61.40 a head.

FIGURE 4.5. Loading Macleay cattle at Cayley, Alberta, 1945. Charlie Glass (left foreground), George Chattaway (right foreground). Photograph property of the Blades and Chattaway families and their descendants.

The next year Macleay sold a similar set of three-year-olds in Great Britain. Though they hit a particularly soft market, he was by then convinced that a long-term reciprocal trade agreement with the Mother Country was worth pursuing. As the chairman of the Council of Western Beef Producers in 1934, he advocated a national cattle marketing plan to the Stevens Commission in the House of Commons, which included lowered freight rates on export shipments and, if necessary, government-imposed minimum prices.[28] He noted that the United Kingdom was the one market that might help the industry. He suggested that Canada increase the preference on British imports in exchange for "an outlet for our cattle." Soon thereafter, however, the United States entered a more liberal period with respect to international trade, and as prices in North America began an extended period of gradual improvement, Macleay and Canadian cattlemen generally were again able to market their product more profitably on their own side of the Atlantic.[29] Evidence suggests that from that time on Rod continued to finish a percentage of his slaughter cattle on grain. Thus, for instance, in her compiled history notes, daughter Dorothy wrote that in 1938 "steers were put in the feedlots at the Bar S, about 350 hd of three year olds."

Evidently, then, Rod Macleay was flexible in his management style and basically prepared to attempt any forms of production and marketing that he felt might help him profit financially. He also understood the importance of hands-on control on his ranch/farms. This was something few other big operators grasped. The great ranches had found it impossible to mould their workforces into a model of efficiency, mainly because the men who actually owned the ranch, or even had a significant financial investment in it, lived offsite—in a number of cases, far away in Montreal or New York or even across the Atlantic in Britain. The Walrond outfit in the Porcupine Hills to the south of the Macleays had an onsite manager who was in charge of daily operations, and it also normally had a couple of foremen, but none of the members of the board of directors or even the general manager or any of the investors ever dwelt on, or even visited, the ranch for any extended periods of time.[30] A similar situation existed on the first Canadian version of the 76 ranch, the original Circle Three, and even the Bar U after 1897 when the owner, George Lane, moved his home to Calgary.[31] Any rancher or farmer today will verify that this is far from ideal, as wage workers on their own will seldom if ever channel their energies toward the success or survival of the business with the same dedication they are able to muster when someone whose personal wealth is at stake is onsite, visible, and firmly holding the reins in his hands.

Rod Macleay could not be everywhere all the time on his vast landholdings, but he developed a management system to make up for that fact. In today's language, he used a management team. First on the team was wife Laura, who looked after the payroll, books, and domestic matters, which included struggling with a never-ending rotation of cooks and, at times, filling in herself. Stewart Riddle (Rod's cousin and brother-in-law) became his assistant manager of operations starting in 1919. He looked after the farming end and the general goings-on at the Bar S. There were also foremen on the Red Deer River ranch and the TL.[32] Rod had various ways of helping such people feel they had a vested interest in his ranches by allowing them the privilege of running cattle, or in Stewart's case, race horses, on his grass. His daughters were also drawn into the inner circle at a very early age. He gave them a cattle brand when they were seven and nine years old.

Macleays relied on a lot of people and a lot of people relied on Macleays. And yet, nobody ever doubted that the old man was the boss to the day he died. This, no doubt, was to some extent a result of the fact that he was prepared to get his own hands dirty. When possible, he helped with the big bi-annual roundups on the wider and more distant ranges himself (and, as we will see, in so doing, sometimes put his life and limbs at risk).[33] He also worked closely with his men attending to a multiplicity of rather manual tasks whenever he could. Following are reports in the *Rocking P Gazette* newspaper of the varied multitude of jobs he attended to, working head and shoulder with men on his payroll over the course of a year and a half in the 1920s.

> "All the calves were dehorned … in the latter part of the month. The job was done at the Bar S, the main cowboys were R. Macleay, S. Riddle, V. Blake, R. Raynor, C. Walters and F. Sharpe;"[34] "The home field was worked by R. Macleay, S. Riddle, C. Walters and Val Blake on May 3rd;"[35] "R. Macleay and C. Walters pulled a cow of[f] the bog at the Calf Camp on the 25th;[36] "Robert Raynor … assisted by R. Macleay and S. Riddle have been very busy lately, building a new hay rack, with which they are going to feed the bulls;"[37] "The first bunch of beef cows were shipped from Cayley on the first. The punchers were R. Macleay, S. Riddle, Bill Kreps, and Bill Livingstone;"[38] "The second bunch of beef was shipped on February 10th. The punchers R. Macleay, S. Riddle and Bill Kreps started them from the Bar S on Feb. 8th. Going as far as the Henry place that night. The next day they made it to Drumhellar's and on the 10th Cayley;"[39] "The calves from Section thirteen were weaned on October 26th by Roderick Macleay, Val Blake, Tex Smith, George Peddie, Max and her "pard;"[40] "On Sunday Oct. 26th, S. Riddle, R. Macleay, R. Raynor, V. Blake, R. Smith, E. Orvis and T. McKinnon figured and figured, pulled, moved and adjusted the brakes on the new grain tanks. They worked all morning and finally

FIGURE 4.6. Talk about getting his own hands dirty, *Rocking P Gazette*, February 1925, 16. Property of the Blades and Chattaway families and their descendants.

came to the conclusion that the rod should be shortened;"[41] "Stewart Riddle and Rod Macleay stooked about twenty acres apiece on Sept 10th (so they say)."[42]

First under Emerson's mentorship, and then on his own, Macleay obviously learned and then mastered all the tools of the trade he needed to run an efficient operation. By the 1930s he was experimenting with some cutting-edge methods for growing wheat as well as barley, he was using selective breeding to improve his cattle, he was feeding grain to some of his steers in a so-called feedlot system, and he was also in the hog and horse businesses. After freeing themselves from various partnerships, the Macleays were still far from alone. They relied on a vast number of people, as he and Laura built a business of their own consisting of the two of them and certain key people, which, later on, they were to expand to include daughters Dorothy and Maxine. That story is critical to our apprehension of the relative efficiencies of the family approach. Accordingly, it is to it that we will now turn.

5

Enlisting the Nuclear Family, 1909–1925

Even more important than diversification, expansion, and hands-on control in sustaining Macleay ranches from 1906 on was the family approach. This corroborates Elliott West's view of the economics of this form of production as the mainstay of settler success in the American West:

> The pioneer household was an economic mechanism of mutually-dependent parts ... a productive unit, often a remarkably effective and self-sustaining one. Fathers did the heaviest labor—sod busting, construction, and fence-building on a homestead ... and took off in search of other wage work when necessary. Mothers handled the multitude of domestic duties, cared for barnyard animals, gardened, and earned cash by washing, cooking, and sewing for others. Children filled in wherever they were needed ... the frontier's popular image is one of individualism and self-reliance ... but the transformation of the nineteenth-century West could be more accurately pictured as a familial conquest, an occupation by tens of thousands of intra-dependent households.[1]

Because the Macleays' holdings were so much larger than average, their division of responsibilities was not exactly as Professor West outlines. However, the fact of the sharing of responsibilities and mutual dependence within the family relationship certainly was. The other person who put heart and soul into the Macleay organization was Roderick's wife Laura. To treat her as some histories of the West have treated women in general—little more than an enigma, a sort of hazy figure hovering in the background—would be unforgivable.[2] Roderick himself gave due credit to Laura as his "right-hand man," acknowledging most of all that no one other than himself played as important a role in the Macleay operation.[3] She was well equipped to help him. At twenty-one she had graduated with a degree in business from Newport Academy and Grade School, with a course load that had included commercial arithmetic, commercial law, English composition, commercial geography, stenography, typewriting, and bookkeeping. As we will see, her willingness to apply her energies, and her education, to work on the ranch provides support for historian Mary Kinnear's argument that women in rural western Canada generally during the interwar period could actually gain something from rising to the difficult challenges of country life. These women, Kinnear believes, appreciated the self-esteem they achieved through their sacrifices and their contributions to the rural economy. They were able to feel that they were true partners on the land.[4] From the time she met Rod in Vermont, Laura Sturtevant's life must often have seemed to her something of a whirlwind. Rod swept her off her feet during that short visit in November 1905. On 12 December, only a few weeks after they had met, the two were married in St. Mark's Episcopal Church, Newport. This was a particularly short engagement for those days, almost scandalous, when a reasonable period was required for just about everything. Like most other things in the Macleays' life, it was motivated by practicalities of time and distance. But it was still quite a plunge, especially for Laura, whose whole lifestyle would change dramatically and irreversibly.

When Laura left her socially refined home in Newport, family and friends predicted that she would last a year at most in the wild and woolly west. In January 1906 she moved with Rod into what by homestead standards was a reasonably nice five-bedroom house then on the home place. But she must surely have experienced cultural shock crowded in with a household of men.[5] Sometime soon after her arrival the men expanded the house with an addition that became a dining room and kitchen. But this could only have provided minor relief to the heavy burden Laura carried in preparing three meals a day for the partners and up to seven other men when the haying crews were at work. Laura no doubt felt isolated in those days as well and, for much of the time, lonely. She did not see another woman for seven months; there were no telephones or gramophones and the post office was eight to nine miles away at the Bar U. If there was time, someone would go to the post office every week, but often two or three weeks would pass before the mail was retrieved.

With her husband's support, however, Laura proved up to the challenge. From her arrival, she took charge of domestic functions. This was a daunting task. To feed the men, she had dried and canned provisions plus flour and other staples hauled in from High River, then the nearest town some twenty-five miles to the northeast, and less often from the village of Cayley (incorporated in 1904), which was about half the distance away in the same general direction.[6] Travel in those days was slow compared to modern times. In the first decades of the twentieth century it was by horse, or horse and buggy, and after the automobile became common in rural districts in the early 1920s, the roads were so poor that it improved very little. Laura and Roderick's grandson explains that even in the early post–World War II period mobility in the Porcupine Hills was still severely limited:

> A good road was any stretch of country with no gates to open and a few people living along the way in case you ran into trouble. A bad road had gates and no hope of help. The road down Happy Valley where Highway 22 [north to High River and Calgary] runs today used to be on the west side of the valley. It was a bad road. It was not actually even a decent trail. Nearly every mile there was a gate and we used to get out and line up the planks for creek crossings.
>
> For weeks we used to compile lists of things we needed, just in case someone went to town. Twice a year Dad would make a major trip to Nanton [just 17 miles to the east] and bring back six months of supplies such as flour, dried fruits, coffee, tea, sugar and assorted dry goods. People used root houses, meat houses, milk cows, chickens and preserves. They were not dependent on how late the "super stores" were open or how well the shelves were stocked. Any trip to

FIGURE 5.1. As this 1924 depiction suggests, after a rainy period, or during spring thaw-up, or when the snow was deep, it was more efficient to use the horse and buggy to transport people and goods than the early motorized vehicles. *Rocking P Gazette,* April 1924, 10. Property of the Blades and Chattaway families and their descendants.

Calgary was an overnight venture and mail was a sought-after treasure. Tobacco and whiskey actually created more trails than they are given credit for and I can't think of a one of us who has to open a gate to get to town today.[7]

Throughout more or less the entirety of Laura's life on the ranch, therefore, when particular provisions ran out or were missed on one of the widely spaced food runs, it was not possible for anyone to make a quick trip to the grocery store. During the winter, she got potatoes and other vegetables from the root house and beef and wild game from the meat/ice house.[8] In the summer months she picked her vegetables from the garden about a mile to the east. In the early years, she worked without even running water. Originally the men carted in spring water in a barrel on a stone boat.[9] In 1909 they drilled a well between the bunkhouse and the blacksmith shop to a depth of 100 feet. It had lots of water, but it had to be pumped by hand. The hole was drilled so crooked it wore out the rods quickly, and in October 1912 they drilled a new well; and then a storage tank, windmill, and waterline provided Laura with the luxury of running water for the first time to cook, launder, and wash.

Laura, and from time to time in the earlier years her cousin-in-law, Margaret Riddle—who would reappear from the East time and again over the next eight years to provide much-appreciated help and companionship—not only coped with cooking, cleaning, and doing laundry but also milked cows, seeded and weeded the garden, and kept, killed, and plucked chickens and gathered eggs. She also learned to contribute to the food supply in a way she had almost certainly never previously foreseen. Margaret appears to have been her teacher in this respect. The pair would take a buggy and a .22 rifle and head out after game. The results were significant. They managed to keep the larder well supplied with sharp-tailed grouse—a welcomed addition to the family diet when quick trips to the store were impossible. Laura became proficient with a gun and she continued to hunt on her own even in later years. A remarkably accomplished poem in the *Rocking P Gazette* of April 1924 was without doubt written to recognize this fact. Titled "The Little Twenty-Two," it was written by Macleay ranch cowpuncher Tommy McKinnon.[10]

FIGURE 5.2. Roderick and Laura Macleay, in 1947, with their grandchildren. From left to right: Betty Blades (standing), Ernest (Mac) Blades (held by Roderick), Rod Blades (standing), Clay Chattaway (held by Laura), and Ethel Blades (standing). Property of the Blades and Chattaway families and their descendants.

> I am going to get married very soon I expect.
> The way we got acquainted was through the Gazette,
> And all that she wrote me all will come true,
> A bunch of prairie chickens and a little twenty-two.
>
> She wrote me a letter so perfect and neat,
> Enclosed with the letter was a photo so sweet,
> She sure looks a real dandy and I hope it is true!
> A bunch of prairie chickens and a little twenty-two.
>
> She calls herself Jessie I won't tell you the rest,
> To keep it a secret I think it is best,
> But when we are married I'll show her to you,
> With a bunch of prairie chickens and a little twenty-two.

Now all of you young fellows a warning do take,
And send in a nice little ad to the Gazette,
And I dearly do hope the same luck will reach you,
And get a nice girl that owns a twenty-two.

Don't get a flapper for you're sure to repent,
Get a nice little girl that is quiet and content,
One that will get out and rustle the dinner for you,
A bunch of prairie chickens and a little twenty-two.

I hope that nice letter wasn't really a fake,
My heart will be broken, my whole life's at stake,
For I am expecting the wedding to go through,
And get that nice girl that owns a twenty-two.

At the end of the poem there is an "Addition" announcing "with deep regret that the girl with the chickens and twenty-two has been wed for years—so this won't come true."

While handling all the tasks noted above, Laura raised and nurtured Dorothy and Maxine. In the remote country environment, this included performing duties the modern reader might overlook. "If you want your hair bobbed, trimmed, plucked, or curled, go to Mrs. R. R. Macleay at the Home Ranch," reads the *Gazette* of September 1924.[11] Over the years, as we will see, the Macleays dramatically expanded their land and livestock holdings, making the employment of a very large labour force necessary. In 1909 there were fourteen men and one woman on the regular payroll; by 1920 there were seventy-three. All of the employees had to be boarded because, as we will see, transportation systems were far too slow to allow men to commute from the nearest town.[12] At that point the Macleays also hired a cook. This was not always a sure thing, however: "<u>FOUND</u>, Amid great rejoicing on the Part of Mrs. R. Macleay and her sister after a strenuous month in the kitchen—one cook," the *Gazette* reported in September 1924.[13] The sister was Gertrude, the widow of Rod's brother Dr. Kenneth. She visited the ranch quite often in the 1920s and stayed for extended periods. The March 1925 *Gazette* reports: "Our eighth cook left on the 28th, Bob Reeves taking him to Cayley. This makes an average

FIGURE 5.3. Daughter Dorothy with the twenty-two and her quarry. *Rocking P Gazette*, 29 April 1924, 29. Photograph property of the Blades and Chattaway families and their descendants.

of about one a month since Charlie Lung left last fall."[14] When a cook was available to help with domestic duties, Laura did not suddenly find time to sit around. She took on extra errands, including making the long and difficult run to High River or even Calgary for parts or equipment that Rod and the men required to keep the ranch running. "Mrs. L. S. Macleay has had a raise this month on account of extra freighting work done for binders and so forth."[15]

Fortunately, as they got older the two Macleay daughters were able to share some of the load. Countless entries in the *Rocking P Gazette* document the full range of their activities at the tender ages of twelve through fifteen.

> "Max and her pard [Dorothy] plucked eleven chickens on Jan. 25th, 1925."[16] "Egg production has increased this month. The first of February was celebrated by everyone having fresh eggs for breakfast."[17]

No doubt the girls sometimes felt stretched by the pull between their barnyard chores and domestic duties on one hand and those they were required to undertake on the ranges with the hired ranch hands and their father, "the Boss," on the other.

> "Bert Beaucook helped by Max and her 'pard' moved 215 head of steers from Section 33 to the Mountain field Sept. 23."[18]

> "Home field worked by the Boss, Max and her 'pard' on Feb 19th. Fifty-six head were cut [out] and then taken over to the Bar S feed ground."[19]

> "R.R. Macleay, Stewart Riddle, Max and her pard worked the lake field on the 27, 28, 29 and 30th of December, cutting all cows and calves and thin cows and heifers."[20]

> "Jan 30th was a very hard day for Clem, Max[ine], and her 'pard'. They worked swift and fast at the Calf Camp separating the fat calves from the beef calves."[21]

Undaunted, the girls at times rode out on their own:

> "Max and her pard rode the west field ... and found 24 more calves that were missed when the field was rounded up earlier in the month."[22]

Both the Macleay girls' cowhand talents and contributions were recognized in a poem in the *Gazette* appropriately titled "The Feminine Cowboy." The author was Robert Raynor, the ranch handyman, who also had the unusual distinction of being a Justice of the Peace.

> See the merry feminine Cow-boy
> As she rides the meadows through,
> Swings her quirt with careless joy,
> While dashing off the dew.

> *Local news.*
>
> Jack Ribordy, Dunk Contrie, The Boss and the two "kids" moved the Beef Cows from the South field to Section seventeen on Nov. 10th.
>
> The Boss, Charlie Walters, Max and her pard. weaned the colts on the 10th.
>
> George Peddie, The Boss and Charlie Walters moved the beef cows from Section 17 to section 23 on the 11th.
>
> J. McKinnon, Alabama and Ed Orvis fixed the Spring at the Hughes place on the 11th.
>
> Bryan arrived at the Bar 3 from Willow Creek on the 7th. He complained of the cold and hit for town on the truck.

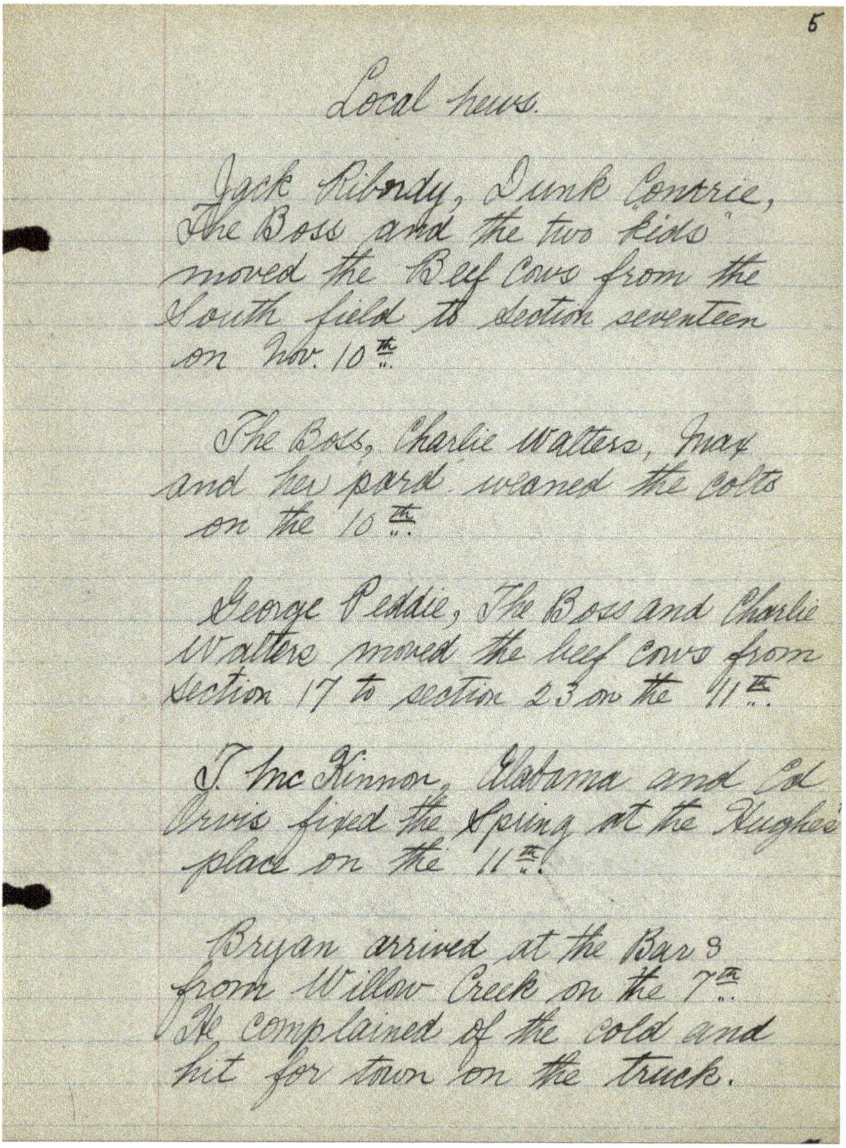

FIGURE 5.4. Dorothy and Maxine lending a hand with and without their father. *Rocking P Gazette*, November 1924, 5. Property of the Blades and Chattaway families and their descendants.

> Riding down the quiet Vale,
> Climbing o'er the hill,
> They differ from the Cowboy male,
> They never stop to roll a pill.[23]
>
> They wear the wide-brim hat,
> And they love to roam
> The range between the U flat
> And the spot called home, sweet home.
>
> They would rather be out riding
> For the Boss of the Anchor P.[24]
> And on the snow be sliding
> Than play golf with their Aun-tee.[25]

As Professor West's description suggests, Laura, Dorothy, and Maxine Macleay's tendency to blur traditional gender roles while working outdoors was a widespread second frontier phenomenon. There are literally hundreds of examples, many in the foothills of Alberta, to illustrate this. On the CC ranch on Mosquito Creek near Nanton, Evelyn Cochrane's responsibility after she arrived from England was planting and nurturing the garden while caring for her children. She too was efficient with a gun and routinely shot prairie chickens and ducks and often defended the chicken coop from "mountain cats."[26] Other women attended to barnyard chores and the care of livestock. In 1901 Katherine Austin joined her husband Fred on their homestead in the Crowsnest Pass area, where they ran both horses and cattle. During that winter, Fred worked out at a lumber company in the Pass while Katherine cared for the baby, looked after their modest home, and fed and nurtured livestock—even donning her husband's clothing so that the milk cow would accept her. It was her milk, butter, and eggs that paid the taxes and much of the regular living expenses.[27] In the same area, Johanne Pedersen was frequently left alone to take care of the family ranch and her seven children while her husband worked as a freighter. Along with attending to her many domestic chores, she was known to "stack hay, stook grain, clear land, saw wood by hand and brand calves."[28] Jessie Louise Bateman of the Jumping Pound district west of Calgary milked cows in an open corral in fair weather and foul. Apparently, she "could milk two cows to anyone else's one."[29] Her neighbour, Susan Copithorne, had come to Canada from Ireland as a child's maid before marrying a rancher. She learned to milk cows, churn butter, and raise chickens.[30]

It was not just on frontiers in western Canada and the United States that this situation occurred. It transpired in the same period wherever frontier conditions prevailed—even on the far side of the globe. At the time the family unit was shaping the cattle industry in western North America, it was also helping to initiate it in three extremely remote and isolated regions of the Northern Territory in Australia. While the climatic and ecological conditions were very different in those regions from those anywhere on the western American plains, the problems of starting a new cattle business where it had never existed before, and where capital and infrastructure were short, was virtually the same.[31] The fact that female input proved just as versatile and as necessary thus helps us emphasize the importance of the latter circumstances. To take just one of numerous examples: near Alice Springs, William Hayes, wife Mary, and six children developed the Owen Springs and Undoolya cattle stations from the late 1870s forward. William reported that "the two girls participated" in running the station, "every bit as much as their … brothers … and they did not hesitate to undertake the same duties." "I understand you acknowledge your daughters to be as good as yourself on the station?" a reporter once asked William. "I do, indeed," he replied, "there is no phase of bush and station life that they are unable to tackle. … They are thorough horsewomen, with or without saddles, and can muster cattle with the best men I ever had." "Can they shoe a horse?" the reporter asked. "Of course they can shoe a horse." "Can they brand cattle?" "Yes; and shoot and dress a beast when the beef has run out. They also break in colts and go out for a week or two at anytime with a couple of [cowhands] mustering cattle. They think nothing of camping under the stars and, in fact, can do anything with stock that men can do."[32]

Collaboration and support by all the members of the family were required to sustain the frontier ranch. In ignoring tradition and working both on the range and in the barnyard, Laura, Dorothy, and Maxine Macleay, like their counterparts living under similar circumstances in other parts of the world, were just doing what they had to do to sustain their way of life. They were called upon to support a system of production and they accepted the challenge because there was little choice. Our next chapter illustrates the contributions Laura in particular made with respect to the strictly business side of cattle ranching. Her input in this area was, if not unique, certainly extraordinary.

6

Finance Matters

Someone asked Rod Macleay in his later years how he had managed to build up and maintain his holdings through difficult periods, which would have included the post–World War I depression and, of course, the "Dirty Thirties." He claimed it was just second nature for him. He said that as a child he had been the little fellow in his family while all four of his brothers were huge, and he had learned the best way to fight was by using his head rather than his brawn. The credit he gave at other times to Laura suggests that he would readily have admitted that her head was instrumental in his success too. Unquestionably, Laura's participation in the Macleays' very complex business affairs was crucial. To clarify those affairs, it is necessary to go back to 1914 and the end of the Macleay–Emerson partnership.

To buy out Emerson, Rod needed to take a line of credit with the Union Bank in High River.[1] Emerson's share of the partnership came to $80,000, and operating capital requirements pushed the loan to $227,000. This was an immense sum for the time, and yet over the next few years Rod threw caution to the wind in expanding his holdings and taking on further and substantial burdens. Arguably, at this stage of his life, still relatively young and certainly ambitious, he made one of the mistakes the great ranchers had frequently made: he took on more than he could handle. Early in the winter of 1916, an attractive ranch southwest of the home place, the TL outfit on Willow Creek, then owned by Dan Riley and his brother-in-law, Fulton Thompson, came up for sale.[2] Macleay wanted it, but his local banker did not have the authority to lend him the money. Undeterred and strong-willed as always, Rod went to Winnipeg and persuaded the "higher ups" in the bank to advance him the necessary credit. On 10 March, he purchased the TL "lock stock and barrel."[3] The total deal was for $92,439. The 1,200 cattle were priced at $55 per head straight through for a total of $66,000; the deeded land, 1,600 acres at $12.50 per acre, cost $20,000, and the lease from the Department of the Interior (#6220) containing 12,878 acres at $.50 an acre, cost $6,439.[4] Macleay paid $2,000 on signing the agreement and another $6,000 soon after. Dan Riley took a first mortgage on the land for $14,000, to be repaid in three equal installments of $4,666 at 6 percent interest starting 1 November 1917.[5] Customarily, the bank took security on the cattle.

In 1918, Rod obtained a permit to graze 1,000 head on White's Creek in the Bow Crow forest reserve not far from the TL. That year he also bought NE 24-16-2-W5 a mile and a half southwest of the home place for $13.00 an acre and the west half and southeast quarter of 19-16-1-W5 for $18.75 an acre. But the major expansion came in 1919, when he bought the Bar S ranch bordering the Rocking P on the south side, from Patrick Burns. This outfit had passed through a number of hands since the turn of the century. Walter Skrine, the original owner, had sold it in 1902 to Pete Muirhead, who sold it to the Vancouver Prince Rupert Ranching Company (VPR) in 1910. In 1917, Patrick Burns, whose father-in-law, Thomas Ellis, was a partner in the VPR, took it over as part of a deal in which he acquired that company's meat-packing plant. When Macleay learned that Burns was prepared to sell, the time seemed right. Cattle prices were relatively high, and there was not another property in the world that could have suited him better: a good set of buildings, some farmland developed, and, best of all, right on the doorstep. Moreover, the Bar S already had a crew in place, and unlike with the other holdings he had purchased, he would not have to construct bunkhouses, fences, and other facilities. It was at this time, too, that Rod's cousin, Stewart Riddle, withdrew from the High River Wheat and Cattle Company. This enabled Macleay to hire his capable and trustworthy relative as his onsite manager or foreman at the Bar S. He knew that, overall, Stewart would take some of the burden in operating what would now be a huge ranching business off his own shoulders. The Bar S consisted of 11,200 acres of deeded and 3,200 acres of leased land, which Rod and Burns priced at $224,000. There were 1,056 head of cattle at the time, which they valued at $90 per head, and 111 horses at $75 per head; the total value of all stock was $103,365. Burns also decided to include 2,280-odd head of mostly big steers running on his Circle Three lease near Macleay's land on the Red Deer River.[6] The price was also to be $90 a head, $205,200 in total, and would have brought the overall cost of the transaction to $308,565 for stock and $532,565 for stock and land. However, this part of the deal was to bring on a major legal dispute between the two men.

The Bar S acquisition was big news right across Canada. An article entitled "A Ranching Success," in the *Review* newspaper at Roblin, Manitoba, read: "The purchase of a cattle ranch of 11,500 acres near High River, Alberta, together with three thousand head of stock for half a million dollars a few days ago was interesting not because of the magnitude of the transaction alone, but because it brought the purchaser into the foreground. This was Roderick R. Macleay, who has long been a prominent rancher in the province."[7] Dorothy Macleay later wrote: "Mom and Dad were thrilled with their new acquisition, this land adjacent to their Home Ranch. Good water, good grass and well kept buildings and all so close!" In the fall of 1919, with the additional Circle steers, Rod made his largest shipment to date. He shipped all these cattle to Clay and Robinson and Miller & Dolan, both agents in Chicago—a total of 114 carloads averaging 19 head per car in six shipments. They brought a total of $262,040. Freight charges from Patricia, on one shipment alone of 38 cars, came to $6,517. Feed, water, and yardage at Moose Jaw was $646. Agents Pendlebury and Maxwell got $47 for issuing the export permits, and C. H. Marshall of Brooks got $75 for supplying hay for the cars. The 2,176 steers brought from $11.25 to $13.50 cwt. for an average price of $120 a head.[8]

Unfortunately, the early 1920s were very difficult in the beef industry and Rod was chronically slow in making further payments. As late as 1924 he still owed Burns $117,337 in principal plus $7,040.22 past due interest from April 1922 to April 1923, as well as interest on past due interest from 1 April 1923.[9] Moreover, the deal brought on a legal battle between the two men, based on an argument over the number of cattle included in the original sale, that eventually threatened to go to the highest court of appeal, which at that time was in England. Beef prices dropped dramatically between the time they set the price in 1919 and the time they were supposed to close the deal. The average price for marketable fat steers fell from $.1306/lb. in 1919 to $.0758/lb. in 1921 to .0675/lb. in 1924.[10] Thus, an animal worth the $90.00 Macleay allegedly had agreed to pay Burns in 1919 dropped to under half that value. For obvious reasons Macleay wanted the number of animals involved in their deal to be as low as possible, and he insisted that he had never agreed to take cattle that were not on the Circle range. Burns claimed that all cattle, including some that were not on the Circle range, were part of their agreement, and when Macleay refused to accept them, Burns sued him.[11]

Macleay hired the future prime minister of Canada, R. B. Bennett, to defend him, and the case went before Judge W. L. Walsh in July and October 1919. Walsh rendered a verdict against Rod on 11 November 1919.[12] At that point Rod instructed Bennett to appeal to the highest court in Alberta and lost a second time.[13] Still he refused to give up. Fearing, and with some justice, that Burns, with all his wealth and political clout, was able to influence the rulings of provincial and even federal courts, Bennett requested and attained permission to bypass the Supreme Court of Canada and go directly to the Privy Council in London.[14] The case was then set for some time in 1921. Macleay must have been worried. According to the previous judgment, should he lose he was liable for $1,100 for every month that Burns had had the animals on his lease and had been forced to see to their care.[15] An unsuccessful court battle that might go on for, say, twenty-four months would have cost him a good deal more than lawyer charges and legal fees.

Before the case went to the Privy Council the two men settled and Burns gave up his suit. Macleay paid for the cattle on the Circle range at the price he had agreed to, but he was vindicated of the claim for the other cattle and all court costs. He also became the owner of the Circle Three brands "0" and "3" that the cattle were carrying. There was every reason for both sides to resolve the issue—Burns was trying to borrow $10 million from banks in New York, which he feared he would not be able to do with a legal suit pending;[16] and Macleay was struggling under what must have seemed insurmountable debt and could not have been comfortable with the prospect of a long, drawn-out court battle no matter how confident he was in his case. By 1924, when he still owed Burns much of the amount noted above, he would owe the Bank of Montreal $459,061.55 and, ostensibly, $25,180.00 in back interest.[17] On top of that, he owed money to the Hudson's Bay Company, the C & E Railway, and the Department of the Interior for land purchased in earlier times, and to various members of his own extended family, including brother Alex, cousin Stewart, Stewart's sister Margaret, and Uncle John Riddle.[18] His total indebtedness had to be in the neighbourhood of $600,000.00.[19] At first glance it seems somewhat surprising the bank did not call in its loan. In 1924, just before the market began to rebound, Rod had 7,889 cattle on the home place that ran from weaned calves to five-year-old steers and mature cows. He also had 338 horses and 51 hogs.[20] An optimistic estimation of the value of the stock would be $325,034.99:

> 1. Slaughter and big feeder cattle: 4- and 5-year-old steers – 1400 lbs. × $.0675 = $94.50 × 810 = $76,545.00. Coming 3-year-old feeders – 1,000 × $.0675 = $67.50 × 970 = $65,475.00. Total: 142,020.00.
>
> 2. Cows, calves, yearlings, heifers and bulls: the Canadian government estimated the average per head value of all beef cattle in the country at $27.11. Since we have taken the most valuable animals out, it would be generous to use that figure for all the rest: 6,109 × $27.11 = $165,614.99.[21]
>
> 3. Horses: 348 × $50.00 = $17,400.00.
>
> 4. Hogs: 51 × $14.00 = $714.00.
>
> Grand Total: 297,934.99 + 17,400.00 + 714.00 = 325,034.99.[22]

The Bank of Montreal was not unaware of the precarious state of the Macleay finances. When Rod asked to borrow another $50,000 to buy stockers in 1923, it refused him based on his indebtedness and his operation's recent lack of profitability.[23] The bank stated that unless Rod could come to some arrangement with Burns and reduce that debt, there was already no way for him to pay back what he owed. It also promised (or threatened) that should he turn another huge loss in 1923 it would review the "whole situation."[24]

Evidently, two central reasons the Bank of Montreal did not foreclose were Rod Macleay's business acumen and Laura Macleay's willingness to work with her husband as a genuine partner even when the road seemed incredibly difficult (and hazardous). Rod understood one fact of business life thoroughly. The chartered banks in those days could only lend money to ranchers on chattel, or liquid assets such as livestock as specified in their loan agreements.[25] Most deals were financed in two ways: the bank funded the cattle and the seller took a mortgage on the land. In the Macleays' case this meant the bank's only security was stock that it had provided money to purchase. From the beginning, therefore, Rod shrewdly and carefully kept any stock he could argue had not been purchased with bank money clearly identified. When he had purchased the CPR lease on the Red Deer River from his partners, he registered the "three walking sticks" brand, as it was called, in Laura's name;[26] when he bought the Circle cattle from Burns he did the same with their brands. This enabled him to feel reasonably confident that Laura's right to cattle so marked would take legal preference over any claim the bank might try to make. Moreover, when Macleay made payments to Burns he sometimes did so in kind, that is, by "selling" him cattle. This allowed the number of cattle the bank could claim to dwindle as the account was paid down.[27] Such a strategy was not ironclad. Had the bank taken him to court and established that its line of credit had been used directly or indirectly to purchase the Bar S, or any other stock, it might well have been able to take some or even all of the cattle. However, it gave the Macleays a very useful line of resistance—one they would utilize for much of the rest of their lives.

In 1923, Rod formulated what turned out to be an ingenious scheme to resolve his debt to Burns. It required the couple to take financial collaboration to a new level, and in the end, it was to be instrumental in keeping the Macleay ranches afloat. At that time the old Gordon, Ironside and Fares firm from Winnipeg, which, along with Burns, had essentially monopolized the western beef trade, was insolvent and selling off its massive leases on the old 76 ranch in southwestern Saskatchewan.[28] In 1923 one of the company partners, William Fares, informed Macleay that a 72,000-acre lease along the White Mud (now Frenchman's) River was for sale. Gordon, Ironside and Fares and Charles Gordon, the son of one of the company's founders, had held the lease and then "assigned" it to the Mule Creek Cattle Company. Robert Gordon Ironside and Charles Frederick Ironside, the two sons of the other GIF founder, were both shareholders in this company.[29] The purpose of the assignment had almost certainly been to keep the lease concealed from the Winnipeg firm's creditors.[30] Fares told Rod that some 1,100 cattle and 35 horses that were still grazing on the land were to be part of the offering along with 140 tons of hay.[31] He indicated that Macleay could have the lease, stock, and feed for the bargain price of $40,000.00. The cattle would be priced at $28.00 apiece—about right on the day's market—but the hay and horses were to be included free of charge. This left the charge for the lease at under $.13 an acre,[32] which was potentially very inexpensive.[33] One supposes that Fares and his associates were prepared to sell at such a low rate for three reasons. Firstly, they could depend on Macleay to keep the deal confidential. Secondly, the term of this particular closed lease had just 4.5 years left, and there was heavy pressure from homesteaders to have all such land thrown open to settlement.[34] Thirdly, buyers were not plentiful at this time because of the depressed beef market.

Macleay realized that the holding could well turn out to be worth far more than the depreciated asking price. When Fares first approached him about the White Mud, he had been actively participating in a leaseholders' lobby effort to get the leases in western prairie Canada allowed much more stable twenty-one-year terms rather than the ten-year terms then in force. He wrote a number of letters to the Department of the Interior, and in early 1924 he personally travelled to the capital for discussions with the Minister of the Interior, the Honourable Charles Stewart, former premier of Alberta, who was naturally sympathetic to westerners.[35] Macleay was sensitive to political matters, and he knew he himself was aligned with other interested and influential parties. His former courtroom antagonist, Patrick Burns, a future Liberal Senator who owned a number of big grazing leases, Dan Riley, a rancher from High River who had sold him the TL ranch and was soon to be a Liberal Senator, and the Western Stock Growers Association of which Riley was president and Macleay an active member (and vice president 1938–39), were also petitioning Ottawa.[36] Eventually the Liberal government bowed to their pressure. In May 1924 Macleay was informed that his own holdings would be renewed for another five years. There was then every reason to believe that the battle for twenty-one years was about to be won.[37] Concrete evidence came in September when he received a letter from Deputy Minister W. W. Cory informing him that three of his current holdings were to be renewed for twenty-one years because they were "located in districts unfit for agricultural purposes."[38] Macleay unquestionably realized that as the beef market improved the much longer terms would dramatically raise the value of such land.[39]

The other important consideration for Macleay was that he could use the White Mud land to further protect his and Laura's liquid inventory—their livestock. He (and she) realized that the grasslands were good enough to carry a lot more livestock than were grazing them at that time. They could fill the lease with two to three times that many cattle branded with Laura's Circle brands and thus insulated by both distance and markings from the scrutinizing eyes and grasping hands of the Bank of Montreal back in Alberta. Their major problem, of course, was that they were not financially in a position to handle this by themselves. Once again, a wealthy partner was required, and the only one available who had something substantial to gain was Patrick Burns. After Macleay visited the Saskatchewan property in June 1924, confirming that its natural pastures were in excellent shape and well watered by the White Mud River, he took a scheme to Burns that he and Laura believed would be good for both parties. It was as follows: If Burns would finance them they would purchase the lease, livestock, and hay and then fill the property with stock branded with cattle legally belonging to Laura. At some stage in the not-too-distant future, hopefully, when the market came back, they would sell the lease and all the cattle to Burns at a friendly price, enabling him to deduct whatever remained of the debt on the Bar S.

That Burns agreed demonstrates that this was attractive to him too—but why? First of all, it would get him paid out for the Bar S. Burns was clearly worried at this time that that deal was at risk. Macleay was in arrears, back interest was accumulating, and he did not want to repossess the property in a depressed market. The Bar S sale had turned out to be a very good one for Burns, and the best possible scenario seemed to be for Macleay to survive financially and live up to the obligations he had assumed prior to the postwar price declines. Burns' papers in the

Glenbow Archives in Calgary reveal that he had actually been trying to get a third party to take over the financing of the Bar S so that he himself could get paid out.[40] Moreover, Burns was not averse to gaining a new grazing property like the White Mud for himself if he could do so at a good price. He had already taken over much of the rest of the 76 land since Gordon, Ironside and Fares had experienced their difficulties, and in a few years he would in fact sell his huge network of packing plants and food wholesale and retail outlets for over $9 million; he would then use his money to buy up and take over indebted ranching properties until he held nearly half a million acres.[41] It could be too that, like Rod, he had assurances that this particular lease would be secure. As a staunch supporter of the Liberal Party he was able to communicate when it suited him with the highest levels of the Mackenzie King government.[42]

So, Burns and the Macleays decided to proceed. Burns initially financed the deal and kept the contract in his own name, but they considered the land and cattle would belong to the Macleays, as long, of course, as they kept up their end of the bargain. They did. By 1928 they had 2,561 head of Laura's cattle grazing the rich grasslands on the Saskatchewan holding along with around 70 horses.[43] There can be very little question they cut back on the number running on the pastures in Alberta. They seem to have begun this process when contemplating the offer from Fares as early as November 1923. According to the *Rocking P Gazette*, at that time the ranch hands had branded "about 450 cows" "with the O."[44] In the Macleay family papers currently on the Bar S ranch there is a typed document by one of Rod's descendants that states as follows: "In 1925 all the Circle 3 cattle, 21 carloads, 819 head mixed were shipped to … Sask. from the home ranch to the 76 range … The herd at home was cut down to 1309 head."[45] The culmination of the deal was at hand. In 1928 Laura got a loan from the Royal Bank of Canada for over $46,000 based on the value of the lease and the stock on it, in order to pay Burns back with interest for his initial loan.[46] Rod also picked up further leases in the White Mud River region from members of the Gordon and Ironside group, the Department of the Interior, the Hudson's Bay Company, and a man named Joseph Kyle. Ultimately, the couple held leases totalling 97,185.23 acres. From Kyle, they also got a section (640 acres) of deeded land.[47]

The deal worked better than either the Macleays or Burns could have hoped. The value of the leases rose as expected, and it just so happened that at the same time the price of beef did a complete turnaround as the postwar depression ended and prosperity returned to the general economy. By 1928 the government's estimated average value of all beef cattle in Canada had risen to just over $57.71 per head, and the per pound price for live beef steers had gone back up from $6.75 cwt. to $10.48 cwt.[48] At that point the Macleays were in a position to sell the White Mud lease and cattle to Burns to the satisfaction of both sides.

The terms of the 1928 agreement between the Macleays and Patrick Burns are preserved in the Glenbow Archives in Calgary.[49] They priced 2,140 of the 2,561 cattle at $70.00 per head, or $149,800.00 in total, and the other 421 head at $50.00 or $21,050.00 in total ($170,850.00 overall). The average, then, was $66.71/head—well over twice what the Macleays paid for the Mule Creek Cattle Company stock in 1924. The leases they put at about $.46 an acre for a total of $45,384.50, which on most of it multiplied the original investment by three and a half times. The Macleays also got $10.00 an acre for the 640 acres of deeded land, or $6,400.00 in total. There had been costs, of course, and death losses among the stock. For instance, not wanting for obvious reasons to bring the Saskatchewan cattle home and not having the infrastructure on the land to fence them into small areas or to supply them with copious amounts of hay, the Macleays had to ignore their own better judgment and take a chance on the weather. The 1927–28 winter was a harsh one and some 205 head of cattle perished during its course alone.[50] Still, the Macleays' position had improved tremendously. The total cost to Burns was $222,634.50. After the Macleays paid out $46,202.00 to the Royal Bank for the money Laura had borrowed, $1,394.40 to the North Scotland Canadian Mortgage Company that Kyle had owed on the deeded section,[51] and $161,634.50 to Burns to settle the original Bar S deal, the Macleays got a paltry $13,797.81 in cash. However, at that point they owned the Bar S unencumbered, and in light of the rebounding economy, their net worth had soared. Their remaining livestock inventory was more valuable than previously, and they still controlled a total of about 50,000 acres of deeded and leased land at the home place, the Bar S, the TL, and White's Creek, as well as 37,000 acres on the Red Deer River.[52]

In February 1929, Burns dutifully informed the Bank of Montreal that he was the owner of all of the cattle in Saskatchewan branded "'O' (circle) left ribs, and/or '3' left shoulder."[53] The main financial challenge for Rod and Laura thereafter was their debt at that institution. Throughout the Depression and World War II period they coped by utilizing the same practices they had adopted earlier to keep the bank's share of their equity as small as possible. In 1930, they formed Macleay Ranches Limited—a family-owned and -operated company, distinct in that sense from the big corporations of the first cattle frontier that had been owned mostly by distant stockholders and operated by hired wage earners—and they put all their landholdings except the Red Deer River property into it. From that point on, Laura wrote the cheques for purchasing replacements when her cattle were marketed, and Rod was careful to see that those animals were visibly identified. We can be fairly sure that Laura's numbers continued to grow as those belonging to the company stagnated or even declined. In 1936 Maxine and Dorothy, now twenty-five and twenty-seven years of age respectively, leased grazing land in their own names on which to run their own cattle.[54] This helped to ensure that whatever happened to their parents' operations they would have assets of their own. When the Bank of Montreal tried to force Rod to give it a blanket mortgage over all Macleay ranch lands and livestock in 1938, he and his lawyers in Bennett's office in Montreal were able to keep them at bay.[55] When the bank brought legal suit against Rod for $370,000 three years later, he filed a defence in the Supreme Court in Calgary contending that the debt was incurred before 1934, and therefore came under the jurisdiction and protection of the Farm Creditors' Arrangement Act of that year.[56] When this was disallowed he settled out of court once again, and the bank reduced its claim well below the amount actually owed, for fear of losing more through foreclosure.[57]

So, did the Macleays use quasi-legal means to make the bank shoulder some of the weight of the debts they had amassed over years while accumulating land and cattle at what might be termed an overly ambitious pace? There is more than one way to look at this. It has been demonstrated in recent times that the financial institutions, many of them from Great Britain, headed out to the Canadian West in the early twentieth century determined to invest huge pools of excess capital in prairie farms and ranches. Mortgage companies, insurance companies, and chartered banks competed feverishly and unrealistically to provide loans to the agricultural sector at interest rates considerably higher than they could have got at home overseas or, indeed, in urban centres of the West.[58] On 7 January 1911, the *Financial Post* reported that the Canadian chartered banks, which would have included the Bank of Montreal, had constructed a total of twenty-six *new* branches in Alberta, sixty-nine in Saskatchewan, and thirteen in Manitoba.[59] At that time 118 chartered bank branches were operating in the three provinces out of a total of 256 in the entire country. The financial institutions' overconfidence regarding the agricultural potential of the West is illustrated too by Rod Macleay's own ability to pile up debts. Their sanguinity proved generally misplaced as the post–World War I depression, the Great Depression, and the droughts of 1916–1926 and the "Dirty Thirties" brought them huge losses.[60] Since the institutions were guilty of assisting farmers and some ranchers in over-investing, it does not seem unreasonable that in cases such as this they were to share the shortfall. We feel obliged to reiterate that one cannot be sure the Macleays' bank would not have been able to take most of the cattle had it decided to pursue them legally. It would have depended on whether it could prove that Rod had used some of the capital it had lent him to buy land or, less likely, stock in Laura's name. However, considering that Rod and Laura had control over their not insignificant part of the paper trail, the bank personnel must have known that that could be a daunting task; and they were no doubt aware that public and media sympathy when such matters go to court is often with the producer. One of the most interesting facts to come out of this episode is the trust and mutual reliance between Rod and Laura as they quietly shifted ownership of their primary liquid resource into her name. It would be going too far to call them equal partners in the ranching business. Rod's reference to Laura as his "right hand man" suggests a close alliance, with him as the senior partner. One could expect little else in this time and place. However, as an expert on male–female associations on the rural western Canadian frontier has argued, "for some women at least, claiming property ownership in the name of family survival could translate into more egalitarian household relations."[61] It seems clear too that Rod and Laura saw eye to eye on business issues. Searching through the family papers one finds no indication of any hesitation on Laura's part over business dealings; and though the Bar S purchase constituted a huge financial burden for the family she, according to Dorothy's report, was as pleased with it as Rod was. Laura had, or at least developed, an authoritative persona in her own right. It was necessary that she have hired help in the kitchen on as close to a permanent basis as possible. It was a full-time job and more than any one person could be expected to handle. She certainly was not hesitant to make her displeasure known when she went without such help for any length of time. She has "been cooking for three weeks" and is "now on the rampage," the *Gazette* noted in September 1924.[62] Reports also indicate that it was part of her responsibility not only to see that dozens of employees were paid but also that supplies were on hand and that everyone was fed.[63] Laura would head off to Calgary herself to find, interview, and hire cooks and bring home the "monthly grub-stake" on a regular basis.

The limiting factors on female autonomy, which historian Dee Garceau suggests the New Woman of the twentieth century sought to overcome, were family authority, domesticity, and female dependence.[64] For Laura, her girls, and many others the ranching experience afforded the chance to rise above all three of those obstacles. Out west they escaped the authority of their Old World traditions and the limitations of a wholly domestic life; and, through their contributions to the family's economy, they overcame the sense of absolute dependence on their male mate. On the second frontier in western Canada, men and women married and produced offspring when their own resources were both limited and being stretched for the purposes of building up their agricultural business; it was essential that they both learn to contribute what they could when they could. This the Macleays had done over a long period. Were there others like them who were willing to use the legally recognized system for establishing ownership of stock—the brand—to loosen or even escape their bank's hold? The answer is very difficult to establish. This is not the sort of thing people normally wanted to talk about, and one very seldom finds reference to it in personal correspondence or business records. All we can say is that, given the freedom that comes with the ability to undertake unreported and unobserved business transactions and to keep chattel in remote locations far away from prying eyes, it must have been tempting. A story recounted in 1905 in the Northern Territory of Australia, by a man working on one of the ranches, illustrates that cattle people everywhere who lived under similar circumstances to the Macleays' could tend to be drawn to similar expediencies. "I was present during the 1900 drought in the taking over by the mortgagees of a station away out," the man explained. "The owner of the property was a married man, and his wife possessed stock in her own name, and these were running on the station with her husband's cattle." As the mortgagee's representative began looking through the cattle he noticed "the station head stockman" was busy cutting out quite a number of the beasts. When he asked what the stockman was doing, "the reply came quick and prompt. 'oh, only cutting out a few of Mrs.'s cattle.'" After they checked all the earmarks and brands it became evident that after the mortgage had been "given over the property, the wife's brand was the only one used on the station."[65] We also know that some grain farmers in western Canada found their own ways to bend legal rules when they apprehended that a mortgage holder was about to seize their land for payment of debts. For instance, some attempted to skirt their financial responsibilities by "selling" their property to their wife. In 1893, one Clara Hicks fought in the county court in Boissevain, Manitoba, to strike off a lien on her farm, which had originally been registered in her husband's name. Since buying the farm from him she had hired her husband to work for her for two dollars a day plus board.[66]

Macleay's bank was eventually mollified. In order to settle with it, Rod sold his beloved Walking Sticks ranch on the Red Deer River, which still left them with a $75,000 deficit to the bank. A farmer and friend named Carl Christensen helped them out with a loan that was repaid in six years. At some stage, they also sold out a share in the "Western Block," on the corner of 9th Avenue and 1st Street West in Calgary, which Rod had bought into in Laura's name way back in 1929. To support the loan from Christensen they gave their friend mortgages on Rocking P land as they consolidated the family operation in the Porcupine Hills. What finally completed the turnaround of Macleay fortunes was the rebounding market. Shortly after World War II, live fat steer prices in Canada rose to $14.63 cwt., as exigencies of war and competition among packing companies like Canada Packers, Swifts, and Burns intensified.[67]

By the end of the war Rod and Laura were in a position to pay off the loan from Christensen and free themselves from debt for the first time. Their land base had been reduced, but for their stage of life and health it was only reasonable. They remained one of the biggest family ranching operations in western Canada. Moreover, the value of their stock continued to climb for most of the rest of their lives, peaking in 1951 just two years before they both died, at $33.50 cwt.[68] In 1953 they left the Rocking P ranch (incorporated 1954) to Dorothy and her husband Ernie Blades and the Bar S ranch (incorporated 1954) to Maxine and her husband George Chattaway.[69] By then both ranches were on firm enough financial ground to withstand disasters such as the foot and mouth epidemic, which sent the cattle industry back into a period of decline even as the two families were taking control. But that is another story.

II.
The *Rocking P Gazette*

7

Introducing the *Rocking P Gazette*

As the reader will no doubt now realize, a significant portion of the information we have used to describe and explain the growth and development of the Macleay enterprises in the Porcupine Hills south of Calgary has been gleaned from the *Rocking P Gazette* newspaper. That fact alone tells us that the paper is a fertile primary source. There is, however, much more in it that speaks volumes not just about this one very well-known ranching outfit but about the social, cultural, and economic attributes of the family operation as it emerged and then in a sense became an institution on the northern Great Plains of North America. As promised, we will now examine the paper closely in an effort to unveil the very large and varied types of information it provides. This chapter introduces the main contributors and provides a basic discussion of the quality and significance of their involvement.

One of the characteristics of the *Rocking P Gazette* that strikes the reader almost immediately is its professional quality, especially considering that during its life, from 1923 to 1925, the two editors were in their early to mid-teens. The writing for the most part is fluent, the grammar precise, and even the spelling remarkably accurate for an age predating computers and spell check. Also, while the overall appeal is unashamedly country and western, virtually every issue features at least one educational article, usually on an Old-World subject such as "Sparta's Bravest Man,"[1] "The Pottery of Peru,"[2] "The Dark Age in Poland," or "Charlemagne."[3] Authorship designated by a pseudonym reveals that Dorothy and/or Maxine were the essayists, presumably using books brought to their attention by their teacher.

FIGURE 7.1. Home schooling for Dorothy and Maxine by teacher Watts. *Rocking P Gazette*, April 1925, 85. Property of the Blades and Chattaway families and their descendants.

The girls were well equipped to edit the *Rocking P Gazette* in part because they were not products of the country school system. On close investigation, historians have discovered that the typical rural school in the 1920s was poorly financed and equipped, with shoddy instruction, often by a single unqualified teacher in a one-room setting where up to a dozen students of various ages and grades had to be accommodated.[4] Though Rod attended meetings of the local Muirhead school board and actually sat as trustee in 1924, he and Laura could not expect their children to begin school three miles away when too young to saddle their own horse, let alone open the many gates; and after the war, the Spanish flu was a real concern.[5] Therefore, between 1916 and 1925, with the exception of two years, they had a hired teacher living on the ranch to work with the girls throughout the school year.[6] If Ms. Ethel Watts, who supervised Dorothy and Maxine's work on the *Rocking P Gazette* (1923–25), is any indication, the Macleays chose their teachers very well.

Fittingly, it was Ms. Watts who recognized the girls' editorial ability, their energy, and their dedication in one of the fine pieces she herself wrote for the paper.

> "Scene – Kitchen"
> Any Evening"
>
> By the Printers' Devil
>
> See them, far into the night,
> Under a dim, religious light,
> Tax their brains and rack their heads
> Till tis time to seek their beds
> For the sake of our Gazette!
>
> See their worried, anxious looks
> Pond'ring deep o'er many books!
> Page after page of pencilled treasure
> They have writ for public pleasure,
> For our "Rocking P. Gazette."

Hush! Hats off to these great minds!
Walk on tiptoe – draw the blinds!
Honour to each lofty brain –
Hard the labour, great the strain,
Producing our Gazette.

One creates deep themes of love;
One portrays the skies above,
One our hearts, with danger, thrills,
One our eyes, with teardrops fills,
By tales in the Gazette.

Tales of knightly deeds, out West,
Filled with song and timely jest;
Days of Ranch and cowboy-life,
Poems of love, and mortal strife
You'll find in our Gazette.

Have you ought to advertise?
Down our columns cast your eyes.
Perchance your needs you'll recognize.
And please, don't harshly criticize
Your "Rocking P Gazette."

When you're far from friends and home,
When in city haunts you roam,
Turn your lonely heart, – peruse.
With home-sick tears, the "local news."
In your Gazette.

Then give Three Cheers for the writers two,
Working by night, and all for you!
May success their labours crown!
May Suns of Glory ne'er go down.
On the name "Macleay," of wide renown
The Authors of our Gazette![7]

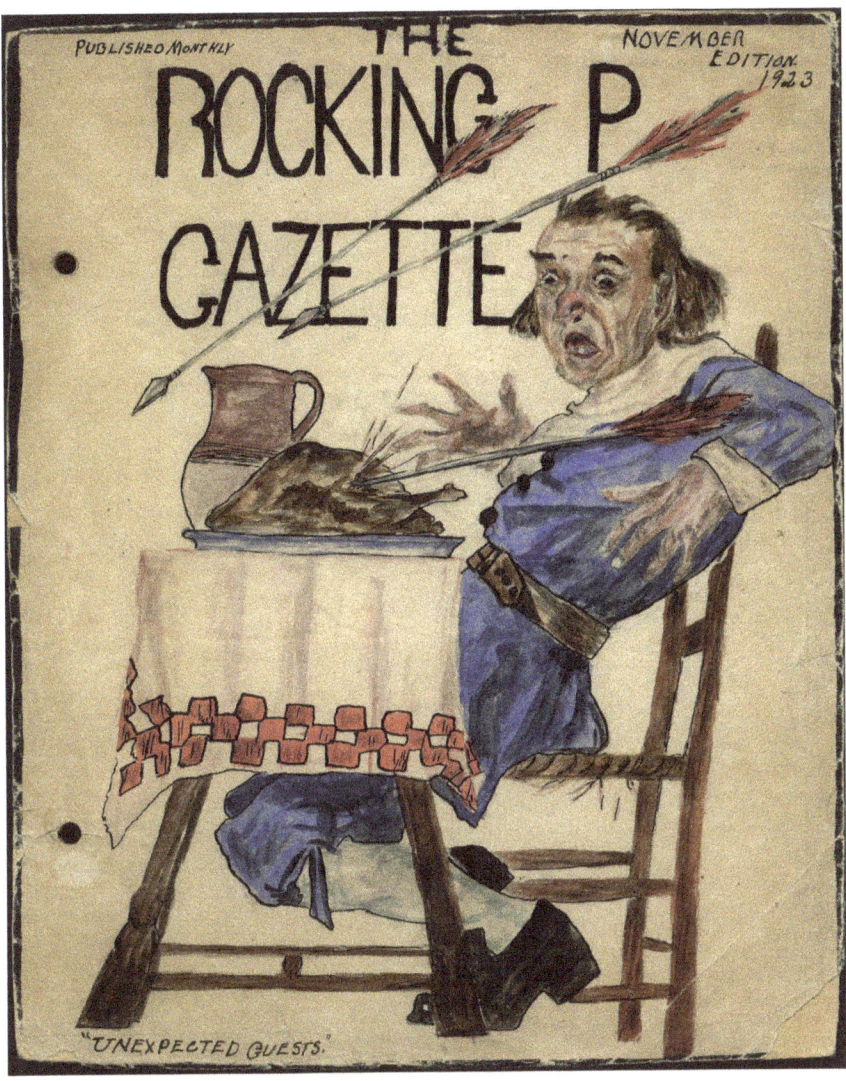

Figure 7.2. "Unexpected Guests," Dorothy Macleay, *Rocking P Gazette,* November 1923, Cover. Property of the Blades and Chattaway families and their descendants.

Figure 7.3. Maxine Macleay's talented depiction of the motion (and the excitement) of "bronco bustin." *Rocking P Gazette*, October 1924, 27. Property of the Blades and Chattaway families and their descendants.

As noted previously, the young editors were also blessed with abundant artistic talents. A number of their own poems and illustrations will be featured here. Above are two of many examples of their visual art found in the *Gazette*.

Some of the plots the young editors presented in their many short stories about second frontier ranchers, ranch hands, outlaws, and relationships are, to be sure, not particularly sophisticated. One argument here, however, is that this reflects their understanding of the types of people they were attempting to reach. From the content, it is evident that the paper was meant for all the men and women who worked on the Rocking P and Bar S Ranches at the time of its production. Because it was handwritten it could only come out in one copy, so after the Macleay household at the Rocking P headquarters read it, it must have been circulated through the bunkhouses on both ranches, where single and relatively young rough-and-tumble cowpunchers predominated. As we will see, while keeping the Macleay household informed and entertained, the paper also attempted to relate to and reflect a bunkhouse culture.

The extent of the two girls' input can be gauged from studying the script. Most of the text is in the handwriting of one or the other. For comparison, easily viewed together are Dorothy's "Cowboy Cal," by "Bucking Barns," and Maxine's "Roaring River Canyon," by "Dan Panhandle," in the October 1923 issue.[8] It would be misleading not to mention that a significant minority of the writing in the *Rocking P Gazette* was done by Ethel Watts, and that she deserves much of the credit for its high standards. She was, after all, responsible in this period for the girls' education. Moreover, she set up each edition by providing the "index" (or table of contents) at the beginning, and she wrote a short story or poem (and sometimes both) for almost every issue, variously endorsed E. W., E. B. W., the "Schoolmarm," or the "Director of Education." Given that she lived with the Macleays at the Rocking P ranch, she was in an excellent position to work with Dorothy and Maxine day in and day out and even when school was not in session. She was able as well to vet each issue of the paper before it came out and to see to the correction of any mistakes she detected or flaws of style, grammar, or spelling. She must also have been a great encouragement to her two charges, in order to keep them committed to the project through seventeen monthly issues.

We have pointed out that women and girls played numerous substantial roles in the early ranching and farming world, and it is fitting that these three particularly capable ladies were able to voice their perspectives through the pages of this remarkable publication. Moreover, that they were able, as will become evident, to get the cowboys and other workers on Macleay ranches regularly to write original poetry for the *Gazette*, and to supply reports about what they considered newsworthy in their working and leisure worlds, brings out the raw, mostly male culture on this one ranch. This is important too. The second cattle ranching frontier in western Canada is distinguishable from the first principally by its longevity (indeed, by all appearances, its permanency). We need to examine it more closely from all its cultural, as well as its agricultural, perspectives in order to gain a better understanding of its attributes—that is, to discover what made it tick. Hopefully, in conjunction with the first portion of our study, the second will help us to take some valuable steps in that direction.

8

The Rural West

A major historical benefit from the *Rocking P Gazette* comes through the authentic glimpses it provides of activities on a developing ranch in the foothills of Alberta. First, as we have seen, it provides a line into the yearly round of ranching activities. These included grain production, the horse business, hogs, and chickens, as well as the vital contributions of women and girls. It also offers information about a host of extraneous events such as the editors' sojourn into the sport of skiing,[1] and golf, the accommodation of the automobile and even the grain truck in the countryside,[2] the advent of the airplane in western skies,[3] and a hunting visit to the Rocking P ranch by Edward Prince of Wales and his stately entourage.[4] Many of these events the editors illustrated with cartoon sketches, which also speak volumes about the world they meant to depict.

Some of the activities the *Gazette* underscores in this way, one might easily overlook. In one issue, for instance, is a depiction noting that two of the ranch hands were busy hauling ice for the ice house, which was used for cold storage of beef and wild game as well as beef and pork. The newspaper shows exactly how it was done.[5]

The *Rocking P Gazette* also represents concepts westerners claimed to hold dear, including freedom, individuality, Mother Nature, and the countryside. At the time, thirteen-year-old Maxine's "Sonnet to the Foothills of the Unspoiled West" projects all these values.

> Between the mountains and the plains they stand,
> The hills where cattle, horses, men roam free,

FIGURE 8.1. The Macleays had their own golf course on the ranch, *Rocking P Gazette*, October 1923, 58. Property of the Blades and Chattaway families and their descendants.

FIGURE 8.2. Ice for the ice house. *Rocking P Gazette*, February 1925, 12. Property of the Blades and Chattaway families and their descendants.

FIGURE 8.3. This block of ice is going to have to be cut up some more. *Rocking P Gazette*, February 1925, 12. Property of the Blades and Chattaway families and their descendants.

8 | The Rural West

> From whose fair heights as far as the eye can see,
> Stretches for aye the boundless prairie land.
>
> Thy rolling grass-lands are by breezes fanned,
> Those breezes clean, and fresh and pure as thee;
> Chinooks in winter come to set us free,
> From the cold grip of winter's icy hand.
>
> This is the land of sunburst sons of toil,
> The land of labour 'neath heaven's wide-spread sky,
> The land of ranges wide, and fruitful soil.
>
> Out in the golden west the old days die,
> But far from the cities' noise and busy turmoil,
> May our foothills still in old-time freedom lie![6]

Embracing environmental conservation to some degree also gave the family outfits an advantage, methodologically speaking, over the earlier company outfits and, along with other factors we have examined, this speaks to their relative longevity. One could argue that the "second frontier" ranchers did more to alter the environment (wells, crops, fences) than the open-range ranchers who essentially replicated the "natural" condition of the area by populating it with cattle and letting them fend for themselves in the same way the buffalo had. However, while the corporations had tended to badly abuse their natural pasturelands, principally by uncontrolled overgrazing, the homesteaders not only cut down their herd numbers to what the land could more realistically be expected to support, they also managed to practise a modicum of rotational grazing simply because the networks of fences they constructed gave them the ability to put their stock on special pastures in the summertime and then to enclose them on land of their own or their neighbours', where feed and shelter were available, during the winter.[7] In the *Rocking P Gazette*, the realization that Nature will provide more when treated with care surfaces in a number of ways. In the story "Forty Years On," Ethel Watts arrives at the home place for a reunion some four decades into the future and is impressed by how little the hills and lakes and animal life have changed in all that time.

"Forty years on!"—"Aye, indeed!", she mused,

"Forty years have seen great changes in the wide world, yet not so hereabouts. Forty years back I roamed these hills with these dear little girls, carefree and happy—now once more we are to meet and renew old ties and friendships!"

She rode on slowly westwards, drinking in the scene, deep in thought. The same old hills looked down, smiling and green, the same sloughs and lakes glimmered in the soft June sunshine. The scene was the same and this June 1st, 1964 it might have been the very same cattle and horses grazing so peacefully in the valleys.[8]

The teacher is particularly appreciative that "the rancher" (now a grandpa), Rod himself, had "through the forty years past ... seen to it that this wide range, though prosperous and flourishing, was unspoiled by the hand of man." In this Watts was recognizing the care Macleay must have taken of his pastures that had been so essential to the survival of the family enterprise.

Another piece that enlists the conservation theme is "The National Parks of Canada" in the January 1924 issue.[9] It acknowledges that "if some steps had not been taken by our government to prevent the total extinction of some of the most widely hunted animals [by creating the parks], we should be without the famous buffalo, whose ancestors roamed these plains, the graceful antelope, the bounding elk and deer, and the sure-footed mountain sheep and goats."[10] These statements might seem odd coming from a ranch publication. As Donald Wetherell has recently pointed out, people on the land tended to view wild animals that interfered with agriculture as enemies that had to be eliminated.[11] In earlier days the cattlemen's stock associations had provided bounties, which they paid mainly to Indigenous hunters, for the extermination of wolves that were wreaking havoc with the cattle and horse herds; and ranchers themselves had used a powerful strychnine poison to eradicate as many of the predators as possible.[12] They also shot cougars and coyotes that threatened the chicken coop.

FIGURE 8.4. The Macleays seem to have been comfortable with hunters on their own land as long as they were after the right type of prey. *Rocking P Gazette*, October 1924, 16. Property of the Blades and Chattaway families and their descendants.

Moreover, like Laura Macleay, many men and women on the land were known to shoot game in order to augment their food supply. However, ungulates—deer, elk and moose—were scarce by 1900. When the buffalo disappeared due largely to the pemmican trade, the wolf lost its food staple and hence nearly wiped out the ungulates, which it preferred to the ranchers' cattle. Eventually, though, the predators were forced to turn to the domesticated herds too. Presumably, farmers and ranchers could support national parks like that at Banff Alberta and the reserve for migrating marsh and water birds north of Last Mountain Lake near Regina (even though they were considered sanctuaries for species that preyed on their cattle as well as on wildlife they liked to eat), as long as the parks were in areas environmentally unsuited to agriculture.[13] In an essay in the January 1925 *Gazette* by a fictitious writer, Beatrice Bumper, titled "In a Thousand Years from Now," the fear that by the end of the millennium "eight generations of elephants, ten generations of whales" and "two generations of giant tortoises" could be "pushed farther and farther back and … finally be left only in Zoo's and menageries" illustrates an environmental consciousness in our two young editors.[14]

Encouragingly, and probably contrary to what most city folks think these days, in the later twentieth and early twenty-first centuries, cattlemen in the foothills have learned to strike a balance with wild animal life even in and around their own grasslands. On the Bar S itself, Rod Macleay's grandson[15] notes that while his mother, Maxine "told of how the sighting of an elk or moose was dinner table conversation when she was growing up," elk are now "common enough to be considered a pest by the ranching community." "It is a matter of conversation," he notes, "if we make a salting trip and do not see a moose." While "grizzly, wolf and cougar sightings are still a matter of conversation … there is lots of it." He concludes that "if competition at the top of the food chain is an indicator we are in step with nature." In "trying to recreate what once was, we are on the right track."[16] No doubt the licensing system to control hunting helps to keep game animals somewhat in balance on the ranchers' pastures today, and the species most threatening to live cattle—the wolf—will in the future be subject to reasonable culls should its resurgence go too far. The key word would seem to be "balance."

FIGURE 8.5. Maxine Macleay's illustration depicting the buffalo as they once roamed the foothills. *Rocking P Gazette*, March 1925, cover. Property of the Blades and Chattaway families and their descendants.

The conservationist ideals expressed in the *Rocking P Gazette* are reinforced by another ideal that flourished in the West in the decades prior to the Great Depression—what historians have labelled the "country life movement." While admiration of country living was not new, it enjoyed a resurgence in the early twentieth century in the hands of the Progressive movement in the United States. In this period people, many of them in urban settings, looked toward country living as a possible counter to the perceived evils of modern city life.[17] As they watched the ongoing rapid growth of cities, and what they saw as the migration of farmers away from the land, they wanted to preserve the essential character of country and the people who dwelt there, before it and they disappeared. This produced what American William L. Bowers identifies as the yeoman myth—the belief that farmers (and by association, ranchers) were an upright, hard-working, law-abiding, and intelligent mainstay of society.[18] They were more moral than anyone elsewhere, principally by virtue of their constant exposure to nature and the land.[19] Somewhat ironically, they also embraced a scientific and mechanized approach to agriculture in a capitalist economy, and they advocated modern appliances in the rural home in order to keep people on the land. In Canada, historian David C. Jones has identified the rural myth deeming farming and country living a morally superior and natural state of man.[20]

This line of deduction is evident in a story in the January 1924 issue of the *Gazette* titled "East is East, and West is West" by "Sixshooter Sam."[21] The story focuses on the evils of life in Chicago by outlining the seamy world of partying and what were sometimes called "sporting girls."[22] The hero is a young man named Dick who at twenty-one is a "large, Handsome, and brave" lad, "a dandy rider, and horse-breaker" and the foreman on a big "ranch near Smoky River in Alberta." One day Dick receives a call from his father, "Old Mr. Simms," to come to Chicago to help him run "a stylish hotel," which is "large and beautiful" and an attraction for people of "any renown." Dick does as requested, and soon proceeds to get involved in the excesses of the city. He is "dazzled by all the young women" and has "several slight love affairs with some of them." His father manages to extricate him from all of these affairs, but Dick grows to like Chicago "and the women." Eventually he falls "desperately in love with a young girl about his own age" whom he courts by taking "her to the shows and other entertainments." When his father intervenes and sends him home, he tells the girl that he is "going back to the West, and that he wants her to marry him" so he can "take her back with him." When she refuses, "Dick, with a heavy heart" asks "another girl and then another but no luck." Heartbroken, he returns to Alberta, where he suddenly sees the error in his ways and vows to spend "the rest of his days" living the unspoiled country life on "the dear old ranch at Smoky River."[23]

One of the facts the *Gazette* demonstrates most clearly is that in keeping with its pride in country life, society in the foothills of Alberta in the 1920s also unashamedly clung to its old-West style and ways. Historically, the best measure of a society comes from the words contemporaries use in describing themselves. In the February 1925 edition of the newspaper, a poem written by one of the more talented ranch hand contributors makes it clear that a cowboy culture still predominated, even on some outfits that had taken up particular farming practices, but that one had now to travel to specific localities to find it.[24] We also incidentally visualize in the poem the growing urban centre of Calgary, now a city of about 65,000.[25] The city had left behind the ranching flavour that had overshadowed all others in an earlier period. Visible too are the cash wheat farming districts on the flat, drier northern Great Plains stretching east from High River all the way across the southern regions of the province of Saskatchewan.[26] The poem demonstrates as well that the cowboy was still a much-admired specimen in the Canadian West; and that by the 1920s Hollywood (the "picture show"), though still to produce the first "talkie," had already assumed much of the role in augmenting his image that dime and romantic novels had played in an earlier epoch.[27]

> WHILE THERE'S LIFE THERE'S HOPE
> When I first came into the West,
> And left the East behind,
> Rough gun-men and cowboys
> There I hoped to find.
>
> I first threshed up near Saskatoon,
> In the province of Saskatchewan,

And found that driving a bundle wagon
Surely was great fun.

But when the harvest was over
I westward still did go,
Looking for the cowboys and gun-men
Like I'd seen in the picture show.

When I arrived in Calgary,
And the snow-capped peaks did see,
I thought surely I must be
In the heart of the Cattle Country.

But no rough cowboys did I see,
Or gun-men brave and bold,
So I hit for the town of High River
Where I'd see them I was told.

I had no sooner come to that town
When my luck came back to me,
For I landed a job on a ranch,
The good old Anchor P [i.e. the Rocking P ranch].[28]

Then R. Macleay the owner,
Over the phone said to me.
 "Now come to the city of Cayley
And I'll send in a team about three."

It was in the city of Cayley
That I first met Stewart Riddle
He was driving a big McLaughlin car,
Through a deep mud puddle.

He told me that Ed, the teamster,
Would meet me down at Kwong's.
So I went down to the Chinaman's,
Singing some old Eastern songs.

And sure enough about three o'Clock
Ed Orvis did appear,[29]
Driving a fiery four-horse team
Which he could scarcely steer

After loading up some heavy salt
And having a little lunch,
We started for the distant Anchor P, –
For Ed he got a hunch

That if we didn't start pretty soon
We would be out of luck,
For then the cook at the Anchor P
Wouldn't give us any chuck

Then followed a long and lonely drive
Up hill and down vale,
Till the sun went down and the moon came up,
To show us the dim trail.

We reached the ranch on the hillside
Just as the clock struck eight,
And Ed he says to me,
 "I thought we would be late."

He introduced me to the bunk-house
And to "Bob," if you please,
But little did I think.
He was a Justice of the Peace.[30]

And at the supper table
Two students did I meet
To tell the truth I didn't know
That western girls could be so sweet.

Then one cold and frosty morn
As the horizon I did scan

I spied a real rough rider
Leading a team with one hand.[31]

He approached at a swift gallop.
And, as he drew near,
I felt a chill go up my spine
And I thought of my mother dear.

He pulled his horse up on its haunches,
In the middle of the yard,
And I thought "now is the time
For me to be on my guard."

He hooked his team to the wagon
Which was loaded up with salt.
And in a tone of stern command
Ordered me to get up on top.

So there I sat beside him
On a hard, cold sack of salt,
While he spat tobacco juice
On every rough spot.

His wild looking bronco
At his side he did lead,
And I thought to myself "wouldn't it be great
If I could ride that stead."

When after a weary mile or two
Along the trail we had passed,
I began to wonder what he would say
If for a ride I asked.

"Now listen, Tex," says I to him,
(For "Tex" was his name),
"What do you say if I should ride
Your horse of roping fame?"

He looked and pawed and shook the ground
I trembled thro' and thro'
At length I looked him in the face
And thought "who cares for you?"

So when he stopped at the Calf Camp
To get a little drink,
I mounted a spirited pony
As quick as you can wink.

Then, seeing the red-headed Stewart,[32]
Riding in the lead,
I galloped on after him
With all possible speed.

Poor "Tex" he was left behind
To go a round-about way,
And you may be sure I hoped
That he would take all day.

Then after riding a mile or two
From a hilltop I did see,
The ranch that I would soon call home
That is the Bar S – y.[33]

Here I met Highland Jim[34]
And Bob and Tom and Dunk,[35]
Then when I was put to herding swine
I thought that I was drunk.

But the greatest shock of all came later,
When I met the terrible "Blake"[36]
For every time he looked at me
My very bones would ache.

For it was rumoured that this very man
Had taken possession of my new shirt

And I was afraid to say anything
For fear I might get hurt.

Then I saw old William Krepps[37]
Whom Blake was wont to shun
Just because old Billy
Was quicker with his gun.

And so the days pass quickly
As I still live in hopes,
That I will soon become a cowboy
Just like the aforesaid folks.

Frank Van Eden[38]

Along with the mention of the picture show in this poem, the *Rocking P Gazette* editors' reference to some of the best-known cowboy actors in the spoof image and caption below tells us that the movie theatre had made its imprint in rural western Canada by the early 1920s. As one film history website reminds us, "The first narrative film—*The Great Train Robbery* produced in 1903 by Edwin S. Porter [had been] a western. Although shot entirely on the East Coast, it contained the essential elements that made the Western a staple of the Hollywood film industry for the next 100 years. Ten minutes in length, it was action-packed with a train robbery, a chase and a final shoot out. It was an enormous success and the most profitable film of its time."[39] In its wake came some 400 "Broncho Billy" films and the first western movie star, Broncho Billy Anderson.[40]

Hoot Gibson was an American rodeo champion, film actor, director, and producer. He appeared in *Pride of the Range* (1910) and *His Only Son* (1912).[41] Tom Mix debuted as Broncho Buster in *Ranch Life in the Great Southwest* (1910) and starred in *The Man From Texas* (1915), *The Heart of Texas Ryan* (1916), *Riders of the Purple Sage* (1925), and *The Great K & A Train Robbery* (1926).[42] Buck Jones first starred in *The Last Straw* in 1920 and by 1925 he had "more than 160 film credits to his name."[43] A number of films actually featured the Canadian West and the Canadian/American borderlands:

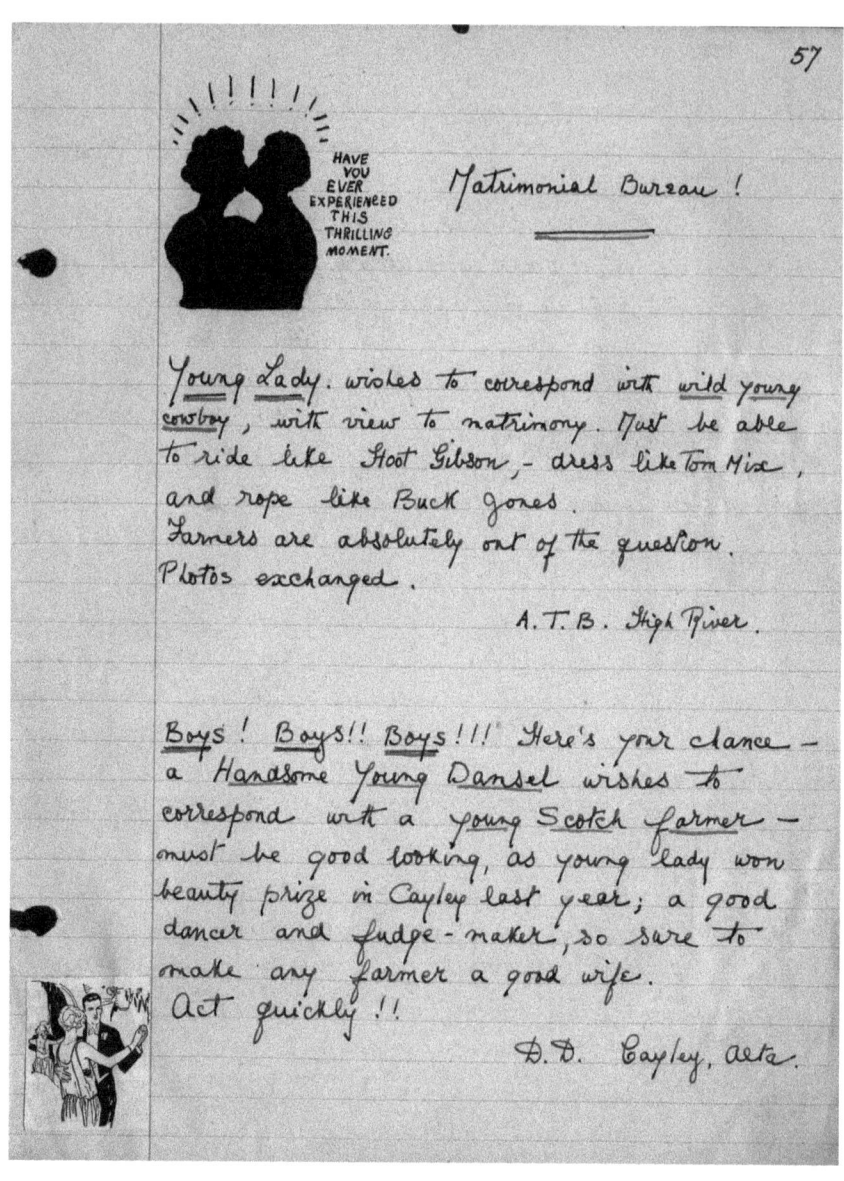

FIGURE 8.6. Dorothy and Maxine and their audience were clearly familiar with cowboy westerns. *Rocking P Gazette*, February 1925, 57. Property of the Blades and Chattaway families and their descendants.

> Beginning with Kalem ... [Company's] *The Cattle Thieves* (1909) in which an Anglo-Canadian Mountie thwarts a half-breed cattle rustler, the film industry used the ranches of the northwest borderlands as the backdrop for cattle-rustling tales typically featuring Indians, Half-breed, or French Canadian thieves. For example, in *The Line Rider* (1914), which takes place in "a treacherous stretch of ground known as Hell's Hole," a Mountie thwarts a pair of cattle rustlers by the name of "Cree Charlie" and Paul Labelle. In *The Half Breed* (1914) a "half-breed" named Moosejaw heads a band of "renegade Indians" in an attempt to steal horses and cattle from the Big U Ranch, possibly a reference to Alberta's famous Bar U Ranch. *Darcy of the Northwest Mounted* (1916) centers on a Mountie who successfully foils Jacques and Batienne, two half-breed cattle rustlers posing as trappers.[44]

By this time, small western towns, including Fort Macleod and High River, had or were getting their own movie theatres, and when the roads were passable the Macleays must have visited one or more of them from time to time, perhaps when Rod and Laura were making a run for provisions.[45] The family also must have read about movies and movie stars when someone brought home newspapers and magazines on one of those runs.[46] Cowboy actors made a major impression in Alberta generally. Hoot Gibson, for instance, made films in both Calgary and High River in 1925. Either way, the film industry clearly helped to augment the stature of the cowboy even in localities where his craft was still in regular use.

To be sure, the major reason, over and above the entertainment media, for the continued strength of the country and western culture in the Canadian foothills was its appropriateness to the job of running the ranch. Despite diversification, the conventions and skills prominent during the days of the open range remained an intricate and substantial part of working the cattle herds. Thus, for instance, a roundup described in the *Gazette* on one of the enclosed leases has all the characteristics of those the cattlemen had undertaken decades earlier. A daunting feature of the roundup had always been the threat of a stampede, which could run thousands of pounds of precious beef off the animals as they

FIGURE 8.7. Poster for Hoot Gibson's movie, "The Calgary Stampede," filmed in Calgary in 1925. Glenbow Archives, Calgary, Poster-126. For other moviemaking ventures in Alberta see, Paul Voisey, *High River and the Times* (Edmonton: University of Alberta Press, 2004), 137–38.

rampaged across the countryside, and bring harm or even death to the men and their mounts as they fought to rein in the unruly beasts. The following rather embellished report of a stampede that occurred in 1923 on the White Mud range indicates that the threat had not gone away as the range was fenced off and divided up.

Arguably, the herd in this case was frightened by the whistle of the train that was to carry them to market; or it might have been trying to flee a very pesky parasite. During the heat of the day, when this event reportedly took place, cattle are their most docile, but the heel-fly (warble) can irritate them endlessly, making them jump and run (usually for the brush) uncontrollably.

FIGURE 8.8. The two cartoons appear with Van Eden's poem. *Rocking P Gazette*, Part A: Tex Smith, February 1925, 26; Part B: Val Blake, February 1925, 29. Property of the Blades and Chattaway families and their descendants.

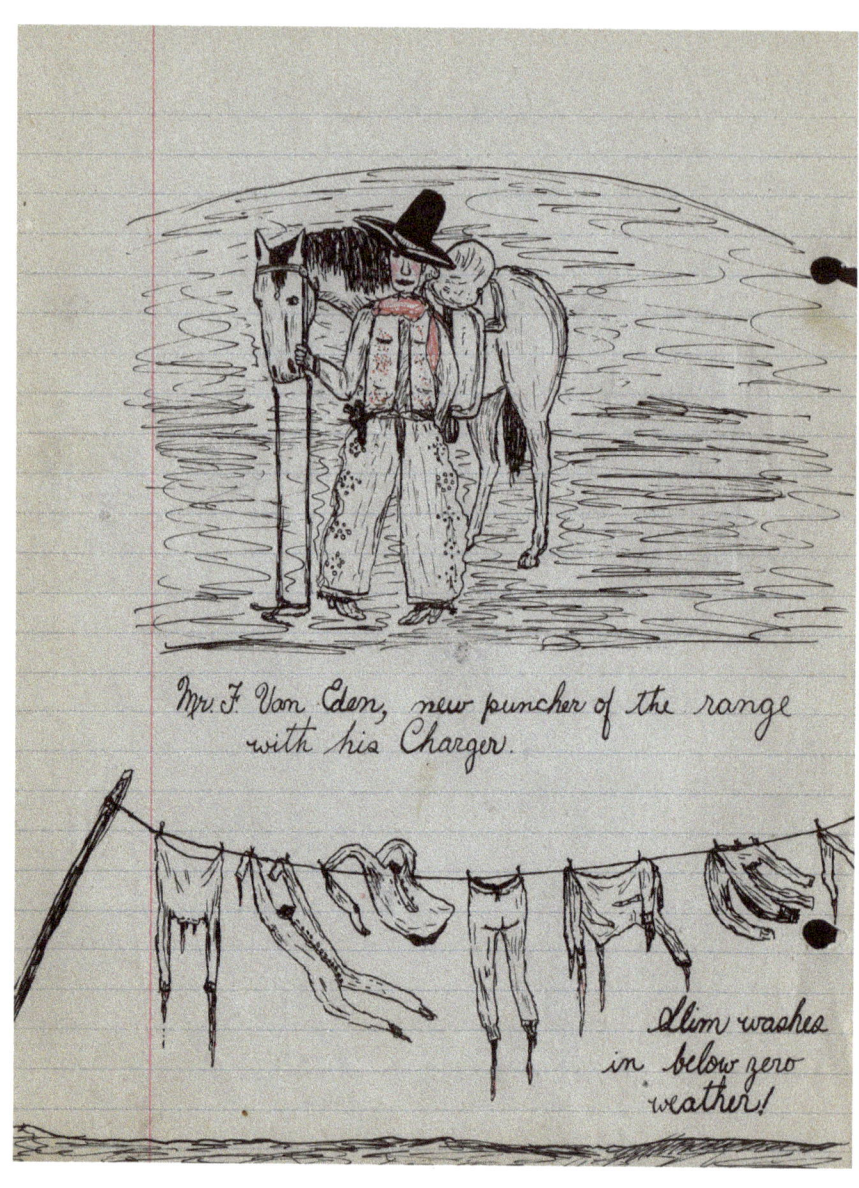

FIGURE 8.9. It appears that Van Eden eventually made the transition to cowboy stature as he hoped he would. *Rocking P Gazette,* February 1925, 14. Property of the Blades and Chattaway families and their descendants.

Rod Macleay, Stewart Riddle, Jack Ribordy, Bill Krepps, Ralph Robertson,[47] Clem Henson,[48] Val Blake, Dair Weise[49] and Wick LeMaster[50] headed along the cattle, while Bill Williams followed along with the horses. And Bill Vesey with the mess-wagon. Everything went well for two days, until Wednesday afternoon. Then:

when everyone was feeling blue with the heat and the herd began to feel gay and stampeded over the hill and through the valley, and up the other side where they met the mess-wagon; The racket scared the team and away it went also, tearing through brush, and open spaces, across creeks, and valleys, and when they boomed across Willow Creek, Bill was all bruised, and shaken up. (But the worst was yet to come.) Bill bounced about two feet off his seat, and, then made a "Swan dive" into the creek, where he lay on his stomach until he saw the herd coming. He had only just enough wind left to paddle across the creek, into the brush where he thought he was safe.

In the distance he could hear the faint cries—"Mill in! Mill in! mill in!." And, back at the herd, the cattle were still gathering speed, on a tin and porcelain path of dishes and … even dried fruit.

Stewart Riddle, Ralph Robertson, and Val Blake were at the lead trying to turn the cattle. Val's horse stepped in a hole, and broke his leg, and got tramped on by nearly one thousand hooves.

As Ralph came tearing past, he stopped and picked up Val, who was pretty badly cut up. He thought that Bill Vesey with his fast team would be at camp, with tent pitched. But, as he neared Willow Creek, Bill emerged from the Willows on hands, and knees, and he was picked up also. A half an hour later Ralph, Bill and Val, came to the buildings where they were greeted by Mrs. Weise, who took tender care of them until late that evening when the rest of the boys came in. They had had a wild ride and several accidents during the afternoon, and at last had succeeded in cornering the herd.

> The next day they started to search for the lost team and wagon, which were soon found, all tangled up in a fence, and from the wagon the mess-box, tent, water-barrel and several beds were gone.
>
> And as for Bill and Val they soon recovered, though Bill had a lump like a hard-boiled egg on the side of his head for the rest of his life.[51]

Because the roundups in big, though now fenced, pastures generally were still central to the operation of the cattle industry, the horse obviously continued to be significant, as did the saddles, ropes, and other equipment along with all the regular skills of the cowboy. Cattle not only had to be mustered, they had to be branded and cut out from the herd when ready for market or to be weaned or sent to the calf camp for feeding. Sometimes, too, animals had to be roped from the saddle out on the pastures so they could be treated for foot rot or pink eye or to have a prolapse or wound stitched up.[52]

> "The Cowpuncher,"
>
> Rides the earth with hoofs of might,
>
> His sharp eye has the grey old eagle's sight
> Where ever he is, he's never in flight
> No matter what happens is always bright
> He'll stay on his horse, no matter how high
> He's there as long, as the sun's in the sky.[53]

The cowboy was very much a part of the second cattle ranching frontier just as he had been of the first. Though he was now forced at times to get out of the saddle to slop pigs, harvest grain, and put up hay, the ranch hand was still required by necessity to ride, rope, bull dog,[54] and brand; and, on bigger ranches at least, he still lived in some dread of the cattle stampede. It was, therefore, out of practicality that the cowboy's legacy endured well past 1912, when the Calgary Stampede erroneously

FIGURE 8.10. These images illustrate some of the many uses for cowboy, horse and rope in the 1920s. *Rocking P Gazette,* Part A: November 1924, 23; Part B: September 1924, 13. Property of the Blades and Chattaway families and their descendants.

celebrated the end of the cowboy era.[55] Evidently, however, and again as in an earlier epoch, it and his country life values were reinforced by media forms emanating to a significant extent from urban and industrial settings. Buffalo Bill and other characters of dime and romantic novels from the late nineteenth century continued to be Western heroes. While by the 1920s the print media Western seems to have been giving way to another genre, the messages it propagated were being preserved and reinforced. We will now attempt to demonstrate another route by which the combination of practical necessity and entertainment value kept a powerful country and western culture radiating from the foothills in the interwar period.

9

Country Entertainment

It is a tribute to the success of the *Rocking P Gazette* in representing germane and authentic country values that a number of the single, young, rough-and-tumble Macleay ranch hands so regularly, and from all appearances, enthusiastically, contributed their poetry to its pages. In this we recognize that they were facilitating and extending an entertainment tradition that stemmed back from the first frontier in the earliest days of the open range. That tradition had first emerged in an oral genre and in the western United States.

When the cowboys guarded livestock at night during the great round-ups in the period before fences, they were often tending to thousands of semi-feral and thus "skittish" animals, even more prone to stampede and on a larger scale than the Macleays' cattle in later years. The men sang to the cattle because they felt it calmed them down. "The singing was suppose[d] to sooth them and it did," one of the cowboys recalled, "I don't know why unless it was that a sound they was used to would keep them from spooking at other noises."[1] The cowboys also sang to pass the long hours when the cattle and the day riders were sleeping and, undoubtedly, to practise for entertaining their friends when they were back on regular duty and all sitting around the campfire at the end of the working day. Some of the men also relied on singing when they were experiencing extended periods of inactivity over the long, cold winters back on the ranch. In Montana, rancher and trail driver Teddy Blue Abbott was a singer, and at the Musselshell bunkhouse during the winter of 1884–85, he used up his repertoire all too quickly. "I knew about ten songs," he said, "and I sung them until everybody was sick of them."[2]

Some songs, such as "The Cowboy's Lament" (derived from an Irish song), "When I Was on Horseback," "O Bury Me Not on the Lone Prairie," or "The Little Black Bull," were well known across the Great Plains from Texas to Canada.[3] Refrains such as "Whoopee Ti Yi Yo, git along little doggies, it's your misfortune and none of my own / Whoopee Ti Yi Yo, git along little doggies for you know Wyoming [or Montana, or Alberta] will be your new home," were repeated over and over again. Some of the songs were improvised and might contain "anything that came into your head."[4] Abbott sang all the old favourites, but he seems to have been particularly proud of his own "Forty Years a Cowpuncher."[5] The song eulogizes the life and adventures of a working cowboy and has a chorus designed to encourage the audience to join in. Most American songs made their way up to Canada, and some came from there. The earliest version of "Blood on the Saddle" was composed at the Cochrane ranch in 1905, and "The Red River Valley" appears to have originated in Manitoba during the first Riel Rebellion.[6]

The themes of cowboys' songs related to all aspects of their life. Among these, along with the stampede, boredom and man's difficult struggle in a harsh environment were mainstays.

> The cowboy's life is a dreary old life,
> All out in the sleet and snow.
> When winter time comes, he begins to think
> Where his summer wages go.[7]

Less musically inclined cowboys who wanted to be artistic would fill long hours, on the range or waiting out winter, memorizing or composing verses. They brought to life the vast range of country poetry, which has been passed down from generation to generation and constantly augmented ever since. While it too naturally tended to glorify men like themselves who were "slow of speech, but quick of hand and keen and true of eye ... wise in the learning of nature's school—the open earth and sky," some of it also dwelt on common emotions about, for instance, "the peril of hoof and horn at the head of the night stampede."[8] Songs and verses rapidly became a prominent feature of cowboy culture, and they travelled with the men once their lonely stints on the range or the bunkhouse ended. One of the reasons they remained popular was that in celebrating the cattleman's life they helped the cowboy maintain the high esteem (and self-esteem) he was never to lose.[9] "While cowpunchers were common men without education," one of them once said, "they set themselves way above other people who, the chances are, were no more common and uneducated than they were."[10]

What their contributions to the *Rocking P Gazette* demonstrate is that the rank and file working cowboys' inclination to produce home-grown country and western entertainment was as pronounced some two to three decades after the end of the open range as it had ever been. As will be evident, it also helps to illustrate that, like similar works in earlier days, this reflected the craft of the cowboy and fed the needs of a remote, socially limited bunkhouse existence. No doubt it signals as well the fact that in the early decades of the twentieth century people from the American West (where the tradition had first flourished) continued to migrate north in larger numbers than from any other non-Canadian country.[11] Of course, in contributing to the ranch newspaper, the Macleay men were for obvious reasons limited to poetry. Three of their poems we have already mentioned: "The Little Twenty-Two" by Tom McKinnon, "The Feminine Cowboy" by Robert Raynor, and "While There's Life There's Hope" by Frank Van Eden.[12] We emphasize that there are a plethora of others in the *Gazette* —"The Wild Buckaroo" by Clem Henson,[13] "The Cowboy and His Dog" by Jim Hendrie,[14] and "The Perfect West" by Ralph Robertson.[15] The vast majority of the poems also in one way or another still replicated or eulogized aspects of the cowboy's life. "Bar S Bill" by Clem Henson gave emphasis to the long-standing view that men of the cowboy profession were better by far than farmers in particular.[16]

> My name is Bill Williams,
> I work for Rod Macleay.
> I've plowed from early morning
> Un-til the close of day.
>
> But now I am a cowboy,
> And over the mountains I will roam.

> Among the wolves and coyotes
> A far away from home.
>
> I'll take my Gibson saddle,
> And old Stony I will tame,
> And I'll defy the Grizzly bear,
> Also the snow and rain.
>
> To—with Jack Ribordy
> His freight teams and his plows,
> For I'm going over the mountains,
> To follow up the cows.[17]

Tom McKinnon's poetic contributions stand out from the rest in terms of style and cadence. In one of his best (to our untrained eye), "Lines to a Wanderer," McKinnon articulated his love for Mother Nature:

> Why, should man seek the glowing west
> And chase the sun to give him rest?
> There he may find Alberta's skies
> Empowering all where beauty lies.
>
> There are no cathedrals rising grand
> To beautify this western land.
> The rockies will guide the wanderer's soul
> But never can he reach his goal.
>
> Alberta, sunny land of the west
> Can give him, more than beauty, rest.
> There woodland aisles o'er shaded throw
> The slanting beams of evening's glow.
>
> The mossy carpet neath the pine,
> Of Alberta's sunny clime,
> Is sweeter far than Roman pride,
> For nature here is ever guide.

> Art may be Art, and for Art's sake,
> There is more within this sylvan brake,
> For Max, the Princess of painters, thought
> When the Rocking P Gazette she wrote:
>
> There is more beneath these western skies
> Than can be seen with naked eyes.
> Yes, I may fly to Western lands forlorn
> But give me aye the place where I was born.[18]

While the *Rocking P Gazette* could obviously not play music in its pages, it did publish cowboy poetry that gave currency to the Macleay cowpunchers' skills in that genre. Some of the boys, it noted, played musical instruments, and one or two fancied himself a singer.[19] Frank Van Eden's "Bar S Nights" evinces that bunkhouse life could still be a challenge when subject to the entertainments of a lone musician. It also shows that in the 1920s boredom among men confined to the same close quarters night after night was as important in stimulating entertainment forms as it had been decades earlier.

> When our daily toils are over
> And our evening chores are done
> We gather in the Bunk house
> To have a little fun.
>
> First we crave a little music
> And Highland Jim is called upon
> To give us a lively time
> On his accordion.
>
> Then no sooner has Jim started
> Than Smokey enters in.
> We know that he can howl,
> But he thinks that he can sing.
>
> But this gets Blake's Irish up,
> For picking up a boot,

FIGURE 9.1. Sometimes the entertainment could get on your nerves. *Rocking P Gazette*, April 1925, 73. Property of the Blades and Chattaway families and their descendants.

"Shut up, you big galoot," he says,
"Or I'll slap your snout."

After listening to Jim's music,
For an hour or more,
We practise acrobatics
In the middle of the floor.

Blake stands on his hands and knees,
Tex upon his head,
While Ted turns somersaults
Over the old bedstead.

But when a poker game is started
Jim and I hit the hay instead,
For we learnt through bitter experience
That we were safer when in bed.[20]

It is testament to just how prolifically this type of entertainment developed out of range and bunkhouse culture that each of the main Macleay ranches could, and did, put together its own band. A cartoon in the May 1924 *Gazette* titled "Bar S Musicians" shows a banjo player, mouth organist, accordionist, and two vocalists playing together.[21] A few months later the *Gazette* presented a program for a "Conservatoire Musicale" at the Rocking P with Ed Orvis playing violin, Dorothy Macleay the piano, Jim Hendrie the accordion, Val Blake the mouth organ, Bill Krepps the "Bones & Banjo," Rod Macleay and Tommy McKinnon the bagpipes, and Laura Macleay and teacher Ethel Watts the "Kazoo." "Maxina Macleay" is listed as the director and "Vocal" trainer.[22]

Another somewhat artistic pastime in which the Macleay men seem to have been almost naturally drawn, and for which their open range predecessors had also been known, was simply telling tall tales. "Speaking of liars," Charlie Russell, who was one of the best, once said, "the Old West could put in its claim for more of 'em than any other land under the sun." This he rightly thought was a reflection of the fact that single men living together would do anything to overcome boredom in the bunkhouse. "They weren't vicious liars," he said. "It was love of romance, lack

FIGURE 9.2. Sometimes everyone in the bunkhouse joined in. *Rocking P Gazette,* May 1924, 10. Property of the Blades and Chattaway families and their descendants.

FIGURE 9.3. The more they talked the greater their exploits. *Rocking P Gazette,* October 1924, 15. Property of the Blades and Chattaway families and their descendants.

of reading matter, and the wish to be entertainin' that makes 'em stretch facts and invent yarns."[23] Apparently, this was still the case in the 1920s. It speaks to the depth of insight of the *Gazette's* editorial staff that they recognized this. Note that the cartoon above is titled "Riding em by the fire."

Hand in glove with cowpuncher entertainments in rural communities from the open range period forward were country dances, often at one of the homes in a specific area, to which neighbours were invited.[24] To a considerable extent because they lived isolated lives, rural people craved and were anxious to take part in social intercourse whenever possible. Once or twice a year they would travel long distances even in bad weather when any kind of a public or private gathering such an agricultural equipment or livestock show or stampede or a musical function transpired.[25] The pages of the *Rocking P Gazette* demonstrate that the Macleays and all their workers at times enjoyed the latter type of event:

> On Feb. 29th a Leap Year dance was held at Mrs. Art Leman's house.[26] Dancing was fast and furious, all the old and new dances being performed by the merry company, with much gusto.
>
> A delightful supper was provided, after which, everybody much enjoyed the impromptu concert given by several of the gentlemen present.
>
> The orchestra consisted of four violins, most ably and untiringly played by Messrs. Charles Waddell, Cecil Lockton, John Hayden and Rooke Herman.[27]
>
> Tex Smith of the Bar S, cut a dashing figure, Tommy McKinnon was one of the shining lights. Mr. T. Johnson and his sons kept their partners hopping while the floor manager, Mr. Jack Smith, kept everybody well up to scratch in the quadrilles.
>
> Art Leman and his wife made an excellent host and hostess. Young Robert charmed the company with his smiles. When the Home Sweet Home Waltz struck up at 3.30 the weary company all agreed that they had had a grand time.[28]

The country and western theme usually prevailed at these events. However, as on the first cattle frontier and on both sides of the Canadian–American border, the British influence randomly also surfaced.[29] In particular the Scottish background of the Macleays, some of their neighbours, and a number of the men who worked on the Rocking P and Bar S ranches is detectable in newspaper reports.[30] Thus:

> On Friday, November 28[th] [1924] a dance was given in the Muirhead s[c]hool-house by Mesdams Comstock, Armstrong and Leman, in honour of the visit of Mr. Albert Comstock. The orchestra consisted of two violins and a guitar, the leading musicians being the brothers Comstock.
>
> During the supper interval the company was entertained by a recitation by Mr. Hadyn,[31] a step-dance and a Scotch song with guitar accompaniment by Albert Comstock.
>
> The Rocking P and the Bar S were represented, Miss E.B. Watts, Tom McKinnon, Jimmy Hendrie, Frank Van Eden and several members of the threshing outfit.[32]

That the Macleay family and their hired help attended such events together may well speak of a levelling effect that derived from the socio-economic realities of life on the second frontier. Two factors underlay this. First, as they worked slowly, and without much certainty, to attain long-term financial sustainability, many of the ranch and farm owners in western Canada between the wars teetered on the edge of insolvency and did not feel there was a very large economic and, therefore, social gap between themselves and their hands.[33] The Macleays were unmistakably among this group, and the fact that Rod actually borrowed money from several family members, including Stewart Riddle, his cousin and foreman on the Bar S, must have prevented this reality from fading in his and Laura's minds.[34] Secondly, many of the men who worked for wages on the ranches and farms in the Canadian West aspired to take out a homestead patent one day themselves and thereby to become a rancher or farmer in their own right. Notable examples are John Ware, who originally worked as a cowhand for the Bar U, and George Lane, who started

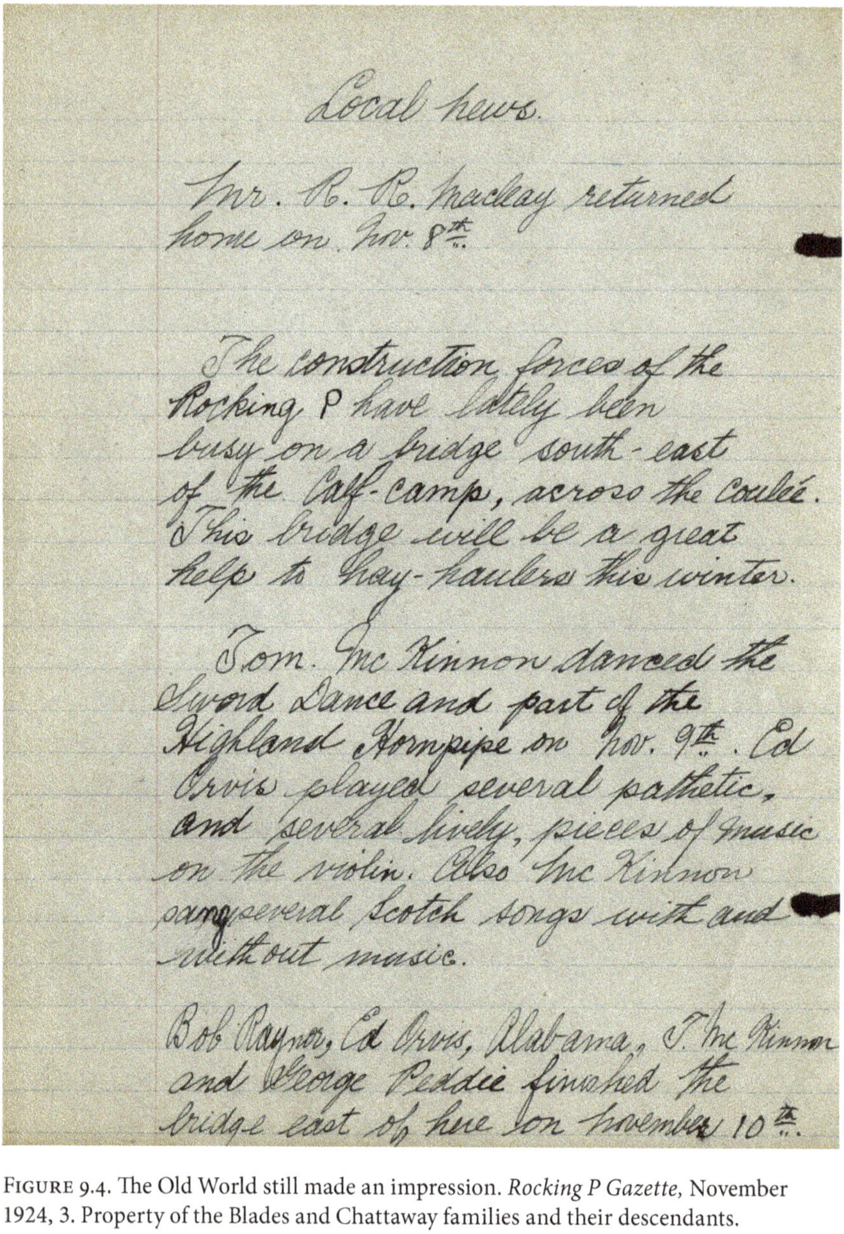

> Local news.
>
> Mr. R. R. Maclay returned home on Nov. 8th.
>
> The construction forces of the Rocking P have lately been busy on a bridge south-east of the Calf-camp, across the coulée. This bridge will be a great help to hay-haulers this winter.
>
> Tom. McKinnon danced the Sword Dance and part of the Highland Hornpipe on Nov. 9th. Ed Orvis played several pathetic, and several lively, pieces of music on the violin. Also McKinnon sang several Scotch songs with and without music.
>
> Bob Raynor, Ed Orvis, Alabama, T. McKinnon and George Peddie finished the bridge east of here on November 10th.

FIGURE 9.4. The Old World still made an impression. *Rocking P Gazette,* November 1924, 3. Property of the Blades and Chattaway families and their descendants.

as the foreman for that outfit and ended up taking it over and greatly expanding it.[35] Another example is Alfred Earnest (A. E.) Cross, who founded the well-known A7 ranch after coming west as one of Senator Matthew Cochrane's employees.[36] How many of the Macleays' workers followed these men's example is unclear, but we know some did. For instance, Donald, Dunk, and Peter Comrie all homesteaded south of the Rocking P and were successful in their own right. Bob Raynor and Hugh Jenkins had a small holding but continued working for Macleays and others. Some left the country, like Charlie Cary, who homesteaded at Bluffton in 1936 and started a place of his own. Sam Howe, Rod's "main man on the Red Deer" River, owned and operated a ranch in that area.[37] The willingness of the Macleay family members to socialize with their workers was also to some degree a result of their need to create a feeling of loyalty and a sense of teamwork among them, which they believed was necessary to provide for ranch sustainability. This need is discussed in chapter 13.

Another vestige of the earlier frontier period in which the Macleay cowpunchers regularly participated, sometimes but certainly not always, with ranch owners was the rodeo or stampede. The tradition of cowboys randomly competing with each other at riding bucking broncos, racing horses, and roping calves started on the southern Great Plains as early as the 1870s and then made its way north.[38] Usually this occurred when several ranches came together for the annual roundups. The *Yellowstone Journal* in Montana commented on one such event in 1885. "The nine-six-nine ranch reports a grand time on Saturday and Sunday at the Capital X Ranch on Mezpah Creek. Over seventy-five cowboys were present and the roping and cutting 'matches' both offered prize money for the winners."[39] There were "race meetings … and, at intervals between races, roping the wild steer, riding the bronco and other events peculiar to a great stock country were indulged in." Well before Guy Weadick, Patrick Burns, A. J. Mclean, George Lane, and A. E. Cross put together the first Calgary Stampede in 1912, professional cowboy athletes had been appearing and competing at such events in the area along with local ranch hands. A North West Mounted Police officer recalled: "the competitors … had often come from a long distance and were past-masters at the

FIGURE 9.5. Rocking P cowboys competed in local competitions when they could. Sketch by Dorothy Macleay, *Rocking P Gazette,* September 1924, 60. Property of the Blades and Chattaway families and their descendants.

FIGURE 9.6. An impromptu bucking event at the ranch. *Rocking P Gazette,* May 1924, 13. Property of the Blades and Chattaway families and their descendants.

games, sometimes champions of the great stock regions south of the line and in our country from the ranches in the vicinity."[40]

Some of the hands working on the Rocking P or the Bar S in the 1920s competed when they could in the Calgary Stampede and in smaller local rodeos such as that at Nanton and at the neighbouring Streeter ranch.[41] The Macleays attended both events as spectators.

Like their first frontier predecessors, the men were inclined as well to hold spontaneous contests of their own. In May 1924 Val Blake "rode a steer at Willow Creek," after the roundup on the TL ranch. The *Gazette* for that month has cartoon illustrations of Jimmy McDonal, Ralph McDonal, and Herb Thurber riding bucking broncos. Thurber is shown inside a corral with people sitting on the rails of the log fence cheering.[42]

As in bygone years, such events were often subject to little or no previous organization:

> It was on a Sunday evening, just as the sun went down
> The Riding Kid of the Rocking P laid his saddle on the ground.
> "Say, boys, I'm going to ride that bronk on Stampede rules, you bet!
> I haven't seen the wall-eyed bronk that ever piled me yet!"
> So the boys they all gathered around the corrall [sic], to see
> That Ridin' Kid from Johnston creek, that rides the Tipton Tree.
> We all got perched upon the rail, to view the Scenes within,
> As Ralph walked out to meet that bronk,
> And gracefully threw his String …

It would be misleading not to recognize that there was an Old World component that also competed with bucking bronco and other stampede events for acceptance in rural western society. In the April 1925 *Gazette* there is a report entitled "Wednesday Night at the Horse Show," which gives a thorough account of an English *equestrian* show jumping competition between "the famous" Calgary horse, Barra Lad, and "the Edmonton horse, Bay Eagle."[43] "Other events … were judging of

FIGURE 9.7. Another impromptu bucking event. *Rocking P Gazette,* May 1924, page 11. Property of the Blades and Chattaway families and their descendants.

FIGURE 9.8. The Macleay cowboys, virtually indistinguishable from their American counterparts. Part A: *Rocking P Gazette,* September 1924, 65. Property of the Blades and Chattaway families and their descendants; Part B: Wyoming Cowboys circa 1915, Glenbow Archives, NA-3466-25.

performance and conformation of saddle horses," including "officers' mounts and polo-ponies" as well as "the show of Bulls and baby beef."

Along with the suitability of cowboy proficiencies to running the ranching industry, the growing American influence helped to ensure that cultural influences from the East and overseas would diminish in the foothills as those of the first cattle frontier flourished. There was one American entertainment component visible in western Canadian society that had nothing whatever to do with working cattle herds. That was the game of baseball. Relatively speaking, it seems to have been an even more prevalent pastime in the 1920s than it is today. The *Rocking P Gazette* indicates in fact that Macleay ranches had its own team. "The Baseball Season has again started," it reported in April 1924. "The Foothill Terrors were out of practice but are gradually warming up to it. The Champion batter so far is Tom McKinnon."[44] Baseball they played primarily for their own homegrown entertainment, unlike polo, which they played in subsequent years (post-*Gazette*) against neighbouring ranches that would gather on Sundays for friendly games. One of the baseball team's major problems seems to have been keeping its membership together particularly during the times of year when matters such as the spring or fall roundups were underway. "'The Foothill Terrors', has suspended playing for an indefinite time, as two of its members and the ball have been lost, "the paper lightheartedly reported in September 1923. It added an "Addendum," that "the ball but not the players have been found."[45] In May 1924 the paper commented: "the Baseball Team has stopped playing on account of the loss of the following players: Tom McKinnon, Jimmy McDonal, Jimmy Hendry and Herb Thurber."[46] All four were hired for their expertise in the cowboy craft.[47]

As historians have demonstrated, baseball swept across the western Canadian prairies as the American population moved in, with many small towns forming their own teams, often even bringing in professional players from across the line in order to compete with other towns.[48] Competition could be so heated that betting, bribery, and even violence sometimes infiltrated the predominantly single, young male communities as a result. The ranch-centred competition seems to have been much more subdued, in part, one assumes, because teams such as the Foothill Terrors represented the family rather than the wider community.

Moreover, the ranches that sponsored the teams were not using them like some of the rural towns did in a desperate effort to boost their image in hopes of becoming the "Chicago of the North."[49]

Evidence in the *Gazette* demonstrates that the Macleay cowboys regularly engaged in an activity that was common not just across the border to the south but virtually everywhere a single male culture flourished—whether in the rural setting, the mining frontiers, or in ocean ports.[50] This was excessive alcohol consumption. In recent years, we have noted that even during the earliest prohibition days prior to 1892, men in the Canadian West found ways to access liquor pretty much whenever they wanted, primarily through the smuggling trade.[51] Traders who understood the demands of a numerically male-dominated society hauled it in from places as close as Fort Benton and as far away as Ontario. Some also made their own booze in illegal stills. In the 1930s, southern Alberta rancher Fred Ings told the story of the time when a whole keg of home brew was left by the cook tent during a foothills roundup. "That was a wild and hilarious branding, and calves that year wore their brands at all angles."[52] The police were powerless to stop the illegal trade and, as predominantly single young men themselves, their hearts were not in it. Sam Steele went through the motions of enforcing the law while he was the commanding officer at Calgary. "The officers and men hated this detestable duty," he said, because it "gave them much trouble and gleams of unpopularity." The prohibition law "made more drunkards than if there had been an open bar and free drinks at every street corner."[53]

In 1924, after the second prohibition period ended in Alberta, some obstacles were placed in the way of excessive alcohol consumption.[54] The United Farmers of Alberta held a plebiscite on its new liquor bill, the most contentious clause of which called for keeping complete control of hard liquor distribution in its own hands. This was seen as a means of limiting accessibility. According to the *Gazette*, folks in the Porcupine Hills joined the majority of Albertans in supporting the clause and thus "moderation."[55] But this does not seem to have had much impact on the Macleay cowboys. Numerous references are made in the *Gazette* to their drinking habits and even to their engaging in the "home brew" business.[56] Drinking then as always was simply a regular part of the unattached man's way of life—as was the hangover.

Sing a Song of Sixpence
Hip pocket full of rye.
4 and 20 Cow punchers
Trying to catch the guy.

4 and 20 Cow punchers
Searching for the guy –
40 and 20 Cowpunchers
Getting awful dry.

4 and 20 Cowpunchers
They have caught the guy
4 and 20 cowpunchers
Loading up on rye.

4 and 20 Cowpunchers
Singing to the sky
4 and 20 Cowpunchers
Looking for a place to lie.

4 and 20 Cowpunchers
All Standing in a line
4 and 20 Cowpunchers
Had to pay a fine.

4 and 20 Cowpunchers'
Heads very sore
4 and 20 Cowpunchers
Going to town no more![57]

In the 1940s, Rod Macleay had to institute a "dry policy" for all employees. He enforced it, too, by firing offenders.

The Macleay cowpunchers tended also to take up an activity that so often seems to have been a part of the drinking young single man's life on frontiers everywhere—blowing their wages on forms of gambling.[58] Ostensibly, some of the Macleay men had done that more than once

FIGURE 9.9. "But when a poker game is started Jim and I hit the hay instead, for we learnt through bitter experience that we were safer when in bed," ("Bar S Nights"). *Rocking P Gazette,* February 1925, 12-13. Property of the Blades and Chattaway families and their descendants.

(see image and caption of figure 9.9) and not just on cards. In May 1924 "a five-dollar race was staged at Willow Creek. The rival horses being Pickles and Cloudy. The latter won easily, and Val Blake collected his five-spot from Bill Young."[59] This too harkened back to the earliest frontier period when casinos were thrown up in a number of western towns offering poker and roulette to men who did not have a family to let down. The same sorts of men also risked their hard-earned pay betting informally with each other on everything from blackjack to horse racing.[60]

The young editors of the *Gazette* were likely not privy to the information, but along with drinking and gambling, of course, young unmarried men also required sexual fulfilment. In the earlier frontier period, prostitutes recognized the demand for their services that the gender imbalance dictated. They flocked into the fledgling towns of southern Alberta and Assiniboia in large numbers,[61] and scores of them "followed the trail herds." They came out west from centres like Omaha, Chicago, and St. Paul and, as Char Smith's work indicates, they tended to move back and forth across the international border as they saw fit.[62] Many travelled regularly between Miles City, Great Falls, Lethbridge, and Calgary to meet the cowboys arriving with the annual supply of slaughter cattle for shipment to eastern markets. Their business flourished and some of the madams who conducted it flourished as well.[63] Lizzie House operated a series of extravagant bordellos in Calgary with furnishings that are supposed to have equalled those of the best houses in the East.[64] Carrie McLean, better known as "Cowboy Jack," started her career as a prostitute in Montana and then moved on to operate her own establishments north of the line. In Lethbridge, she had "an imposing two-story house with a horse trough in front where drunken cowboys frequently dumped their frolicking pals."[65]

By the 1920s more married couples and families had arrived and, therefore, the trade seems to have kept a lower profile in the smaller towns than it had earlier. However, in larger centres such as Lethbridge and Calgary, it continued to flourish and, seemingly in keeping with the fact that younger males were still numerically ascendant, it managed to operate on a fairly large scale. "According to representations made to [this newspaper]," the *Lethbridge Herald* reported in 1923, "the city is run wide open. A large number of houses in the downtown quarter are openly selling liquor, and many of them are houses of prostitution." We are told that "women only solicit men in the streets and from the windows and [verandas] of houses, where they are seen in half-dressed condition. It is urged that the city authorities and the police are not doing their duty in the way of enforcing the law in these matters."[66]

The second cattle ranching frontier was, then, a society with two basic components. On one hand, it was composed of family-operated ranches whose owners recognized the need to maintain reasonably close relations with each other. Sometimes they worked together to put up hay, feed and pasture cattle in the wintertime, or run the local one-room school; and they also played together on special occasions when enough of them could drop what they were doing and make the trip to a local facility or to a hospitable neighbour's place for an evening of singing and dancing to the tunes of a country orchestra that, likely as not, was made up of home-grown talent. On the other hand, this was still a society where the single man was well represented. While occasionally attending social functions with, and even entertaining, one or more of the local family groups, that man also was apt to sing and play musical instruments for his peers, race and bet on horses or cards, "rodeo," make his own illegal whisky, and sometimes drink himself into oblivion. Though he held dear the country life philosophy of the families around him he, like the fictitious foreman of a non-existent ranch near a non-existent Smoky River, might also at times be tempted by the decadent urban world to extend participation in a social life particularly suited to his specific needs. "Mr William Kreps has left the snow covered pinnacles for the citie's [sic] bright lights and exciting life," the *Gazette* reported one winter. "He will reside at the Empire Hotel [in Calgary] … until spring thaws this land of ice and snow."[67]

10

Principles of Need

As historian Mary Neth has illustrated, the American midwest was won in no small part through co-operation between genders and neighbours in agricultural communities during the earlier years of settlement when cash was scarce and some of the infrastructure of a more developed society as yet unavailable.[1] We can make the same observations with respect to the rancher/farmers in the Porcupine Hills of Alberta in the early twentieth century. By the time they split with Emerson, the Macleays themselves were constantly short of money, and though they had infrastructure such as barns, corrals, and many of the multitude of fences required for ranch operation in place, they and all their neighbours still lacked efficient transportation facilities. The roads, as we have seen, were so bad as to make it impossible to move services or labour quickly in or out of the area.

For that reason alone, cohesion and co-operation among district men, women, and families were required not only to handle a multitude of tasks for building, maintaining, and conducting their businesses and building their community but also for confronting emergencies of one form or another. One of the best indications of this is found in a report of a grass fire in the April 1925 issue of the *Gazette*.[2] A professional fire-fighting facility was another type of infrastructure Macleays and their neighbours lacked and, given the roads, they could hardly hope to bring one in as necessary from one of the bigger, not-too-distant towns like High River, and certainly not from Calgary.[3] Rod Macleay himself was in Calgary when this fire was first reported, and Laura must have notified him by telephone, which by then was servicing most of rural Alberta as well as the towns and cities.[4] Macleay had to take the C & E train to Nanton, where Laura picked him up to get him to the "old shack which was as near as the car could take him" to the scene of the conflagration.[5] Following is the report:

> A short time after noon on Thursday, April 30th. A large cloud of smoke was seen in the South West.
>
> Val Blake galloped post haste to the top of Muirhead hill to investigate. About four o'clock he came back … with the news that the fire was burning from Thorpe and Cartwright's into the Half Way place.[6]
>
> Cowboys and farmers were summoned from their work to the scene of action—they hurried on horseback and with teams into the hills, leaving Frank Sharpe fuming and fretting with a broken bone, in the bunkhouse.[7] Very soon the whole country-side was aroused by Stewart Riddle on the phone and cars, tractors, ploughs, trucks, double-wagons, 4-horse teams and Fords came to swell the ranks of the fire-fighters. …
>
> All that night they fought the fire backed by a very strong west wind. When the morning of May 1st dawned, it had been conquered in most directions, but at one point was burning more furiously than ever and the fight was not won till the day was well spent. …
>
> Great work was done by the men who worked strenuously, some for 24 hours without a rest and also by the Fordson tractor owned by R. L. McMillan.[8] It ploughed valiantly through the night, cutting up miles of fire-guard. …
>
> Clem Henson and Hugh Jenkins did good work with the coffee pot at the Thurber shack.[9] The new Provincial Police-man from Nanton worked through the night and well into the day. Tex Smith fought so furiously that his feet swelled beyond the dimensions of his riding boots, which he had to dis-card and hang on his saddle.

Women and girls answered the call too. "We can't say how many loaves of bread and pounds of coffee were consumed but we are sure they were no more than the valiant fighters desired and deserved. Mrs Martin and

Miss Martin worked all night with Mrs Macleay (school ma'am hovering in the background) to feed the hungry."[10]

The fire was deadly and it caused a lot of damage, but without the action by virtually everyone, including members the local Indigenous nation, it could obviously have been a lot worse: "Thorpe and Cartwright lost all their winter feed and all their corrals and buildings, but were lucky enough to save their house around which Pete Comrie and one Indian fought until they had to ride for their lives.[11] Gardiner lost several stacks and his hay flat was burned out."[12] Though Macleay's "Half Way Place" was damaged, fortunately the best grass land and the cabin were saved.[13] In giving expression to its pride in those who "willingly turned out to fight the" flames, the *Gazette* quite rightly noted that this was "everybody's fire."

This, then, was a case of people recognizing the basic requirements of self-preservation. An unrestrained blaze could sweep across the dry pastures and skip over the diminutive trail-like roads with a vengeance. In earlier days when the country had been far less densely populated than in the 1920s, fires had often been much less well attended and they had at times caused a lot more damage than this one. In 1901, for example, a fire had broken out near the town of Gleichen when a man named Dan McNelly carelessly threw a match into the grass after lighting his pipe. The flames quickly spread over some fifty square miles, destroying the precious grasslands and causing horrible depredations among the livestock. "Hundreds of cattle, horses and wild animals … perished outright or were so badly marred they had to be killed or dragged themselves off to some secret spot to die lingeringly. Whole bands of horses and cattle were burned to death in the bottoms of the Little Bow [River] … many being still alive when the range riders found them after the fire and mercifully shot them down in scores."[14]

By the 1920s the *rural* western plains were as densely settled as they would ever be. It was at that stage that people could come together in sufficient numbers to deal effectively with such a threat. Episodes such as this help us understand other characteristics about rural western society during the second frontier. These include politics. Frederick Jackson Turner argued that the frontier had a democratizing effect on American society because it stimulated reliance on individual action and initiative among its participants as they strove to build new homes and farms out of the wilderness. This event suggests that a similar—though slightly modified—interpretation could be applied to ranching society in the western Canadian foothills. As people faced together the challenges associated with life in a remote region with very limited infrastructure, and as they co-operated to take on local emergencies as well as a host of other undertakings necessary to sustaining rural life, they learned reliance on their collective self, which in turn gave them a sense of the importance and potential of citizen action. How quickly the farmers, ranchers, and hired help organized in this instance and how readily they assumed responsibility for their own welfare suggests that by 1925 they had thoroughly embraced that concept. There is no reason to believe that the same could not be said for regions throughout the rural Canadian West wherever conditions similar to those in the Porcupine Hills prevailed. This event provides context from which to view the rise of farmers' co-operative movements in the early twentieth century, including the United Grain Growers, the Alberta Farmers' Co-operative Elevator Company, the Manitoba Grain Growers' Association, and the Saskatchewan Grain Growers' Association.[15] It also helps one comprehend how a farmer and rancher's lobby group turned political party, the United Farmers of Alberta, could sweep into power in 1921 with a platform based on democratic and citizen enlisting principles, including "Proportional Representation of All Classes," "Direct Legislation and Recall," "Adequate Notice of Election," and "Abolition of Patronage," as well as "Encouragement of Co-operation."[16]

We would argue further that the same sorts of need encouraged rural rancher/farmers to overcome some traditional biases with respect to both gender and race. Albertans generally have at times been described as racists, and they have been accused of upholding a traditional, stereotypical and rather chauvinistic view of masculinity.[17] Some of these depictions have come from the hard-nosed research of modern historians looking back on the frontier period.[18] The argument here, however, is that the *Rocking P Gazette* stands as evidence that such portrayals could be grossly overstated in reference to ranching society in the foothills in the 1920s.

FIGURE 10.1. Neighbouring ranchers cooperated to take on this challenge too. Dorothy Macleay's depiction of a Cattle rustler. *Rocking P Gazette*, May 1925, cover. Property of the Blades and Chattaway families and their descendants.

Dealing first with the gender issue: there was in fact a tendency among both sexes to allot a relatively high stature for this time in history to women and girls that stemmed first from the blurring of gender roles. As we have seen, frontier exigencies forced men and women to work closely together in pursuing and defending their operations and, as Laura's example demonstrates, this could include business. This brought a reassessment of female capabilities, which can be detected in some of the short stories the two young editors featured in their newspaper. A good example is a piece Maxine wrote under the pseudonym "Carney Mulligan," entitled "Canyon Callum."[19] In this story, rancher Mike Callum and an outlaw named Dead Shot Dan get into a gunfight over Mike's daughter, Canyon, whom Dead Shot covets. Dead Shot is faster on the draw and he gets off the first shot, mortally wounding Mike. The assertive Canyon does not hesitate to take revenge: "at almost the same second … another shot rang out, 'Dead Shot' fell forward on his face dead" and "out … stepped Canyon with a smoking colt in her hand." After her father dies from his wounds, Canyon, "now the boss of the ranch," immediately takes over and makes changes necessary to the future. She appoints "one of the boys, Starr Skinner, foreman, firing Shorty McMillan the old foreman, because he has fallen in love with her." From that point "everything went well for a year, then the boss and the foreman joined to run the [ranch] together." Even then Canyon is not content suddenly to become the doting wife dutifully limiting herself to domestic duties. While she gives up "shooting men" and "outlaws," she keeps "up her target practice and riding" and continues to be a major player—a force to be reckoned with—in the world outside the home.[20]

Interestingly, this story could be seen as foreshadowing the young editors' own futures, as they were both to marry and bring in a husband to help them run their respective inherited portion of Macleay ranches. They were both also to be forceful in business as well as operational matters.[21]

As some of the poetry such as "The Little Twenty-two" and "The Feminine Cowboy" demonstrate, the high estimation of the abilities of women represented in the "Canyon Mulligan" story would not have been disputed by some in the *Gazette*'s male audience. We would argue that that opinion was widespread on the ranching frontier as men recognized women's crucial contributions.[22] The other factor that gave frontier

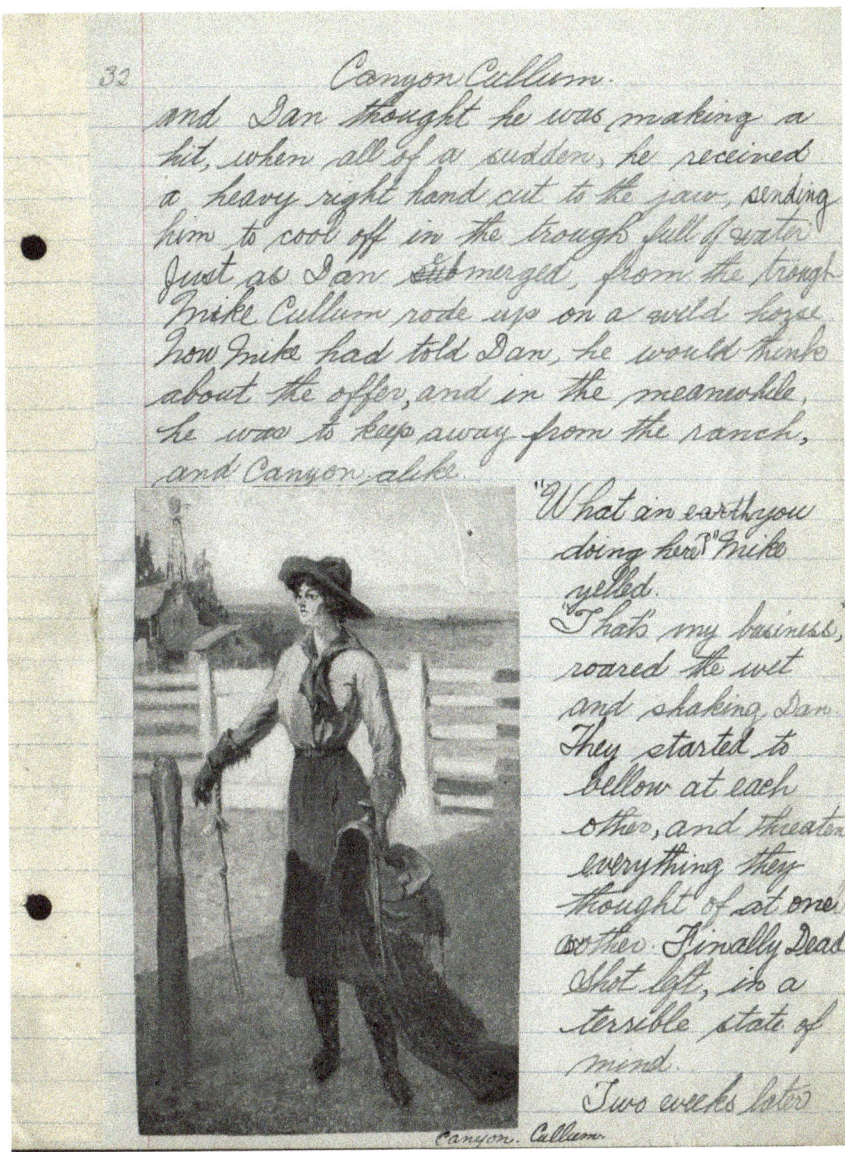

FIGURE 10.2. Canyon portrayed as an accomplished outdoors woman ready to saddle her horse herself. The whip suggests she is not an overly gentle rider. *Rocking P Gazette*, September 1923, 32. Property of the Blades and Chattaway families and their descendants.

FIGURE 10.3. Gender roles were blurred in leisure activities too. *Rocking P Gazette*, April 1924, 9. Property of the Blades and Chattaway families and their descendants.

women a power of their own was simply their small numbers relative to males. The *Rocking P Gazette* represents a society in some ways still proudly tied to the past, and it also reflects the reality that in one important respect that society had not changed a great deal demographically. In the earliest days of settlement, around the turn of the twentieth century, males had outnumbered females by two or three to one depending on the geographic location.[23] For that reason young men had been constantly preoccupied with finding a mate. By the 1920s the gender gap in Alberta had narrowed, but it had not by any stretch disappeared. In 1921, there were 324,208 males and 264,246 females in the province. The disparity was about 22 percent. However, in the ages where both genders might be expected to be most interested in matrimony—between 25 and 54 years—there were 142,741 males and only 98,568 females.[24] The disparity was almost 45 percent. Moreover, in the countryside, where the wage-earning component on the bigger farms and ranches was overwhelmingly male, it had to be considerably larger than in other more populous areas.

In this social environment, young frontier men met few women, single or married, in the course of life, and for that reason alone they had been liable to feel uncomfortable and somewhat inadequate in their presence. "I can't remember that I ever spoke to but three good women in all the time after I left my family," a Montana cowboy reminisced, "and they were all older women, or at least they were married … I'd been … living with men. We didn't consider we were fit to associate with [a good woman] … the cowpunchers was afraid [of them]. "We were so damned scared for fear that we would do or say something wrong—mention a leg or something like that would send them up in the air."[25] One cowpuncher related an incident involving a school teacher who "was a great favorite with everybody." One evening when he was at a dance the floor manager announced a "Ladies' Choice." He "heard that call and figured [he] was out for that dance—and took a big chew of tobacco." "To my surprise," he said, "this little lady stepped up to me and asked me for that dance. Now I had no chance to get rid of that chew and rather than let the little queen know I chewed tobacco or lose that dance, I swallowed the whole works, tobacco juice and all." Later, the same guy commented on the "high regard and respect we had for those good women of that day, as we saw so few of them." He told about an "old hard-faced cowpuncher that had a grouch about something," and would be spouting off about it. "When one of those women would give him some little attention, his face would soften up until you couldn't tell it from the face of the Virgin Mary."[26]

Dorothy and Maxine understood how relatively rare, cherished, and sought after "good women" still were among the predominantly single males on the Macleay holdings, and in issue after issue of their newspaper they gave vent to what they saw as their perpetual quest. "Special notice will be called to the cowpunchers, who since the new 'school marm' has arrived [for the Muirhead school], are looking fine in new chaps, spurs, shirts, etc. and in some cases even new overalls," they jokingly commented in their September 1923 issue.[27] Months later Robert Raynor reported for the *Gazette*:

> The annual meeting of the Muirhead School D[istrict] # 2032 was held in the Muirhead school at 2 O'Clock P.M. on January 12th. Ratepayers present were; P. Comrie, T. W. McKinnon, A. Leman, Chas Dew, H. Jenkins and R. Raynor. A. Leman was re-elected trustee.[28] It is amusing! You can hardly get a connected sentence out of those dry-hide bachelors when it comes to educating the children of the district, but it is surprising, it is marvelous, how they emerge from their semi-coma state when the subject changes to whether we have to engage a man or a lady teacher. The tidal wave was running heavy in favor of a lady teacher.[29]

The girls realized that their ranch hands envisaged forming a relationship with any female of the right age who happened to appear in the countryside. They may have known, too, that few suitable ladies ever settled very long in any of the districts without being bombarded with proposals for marriage.[30] Frederick Ings thought this worthy of mention when he composed his memoirs in the 1930s. "In frontier countries, girls are scarce, and so it was here. Hardly had a visiting sister, niece or friend arrived, than she was besieged by suitors. Practically, every girl or young woman who came in married at once. In fact, it was looked upon as a for[e]gone conclusion."[31] Ethel Watts, herself, was engaged to Tom McKinnon, one of the Macleays favourite ranch hands, just in time for it to be mentioned in the later editions of the *Gazette*.[32]

It also reflects their understanding of their audience that on average the girls wrote at least one short story for each issue about one or more cowboys searching or competing for a country damsel. All the stories offer refreshingly unencumbered plots. "The Easter Lily" by "Coyote Cal" in March 1925 relates how Pete the cowboy, after inflicting a beating on another potential suitor from the city, finds the power to express his love to the young lady named Lily through a flower of the same name at Eastertime.[33] "Mixed Up," in September 1923, is about a detective named Slim who, after falling in love with a cowhand named Curly while the latter is disguised as a woman, in the end finds true love with Curly's sister Shannon.[34] In "The Dying Cowboy," by "Antelope Al," a ranch hand bonds with a young lady after being fatally hurt in a fall from a wild bronco. "Ann leaned over and met the lips of the dying cowboy. His brown eyes smiled up into hers for an instant, his grey lips twisted and he passed over the great divide."[35]

It also reflects demographic conditions that near the end of each monthly issue of the *Gazette* Dorothy and Maxine included a personal column designated "matrimonial bureau." Beyond doubt, it was meant as a spoof, but it would be difficult to make the argument that it was socially irrelevant.

These entries come from the December 1923 issue.[36]

> Handsome young lady wishes to correspond with attractive Cow-puncher with view to matrimony. Good flap-jack thrower and whistler. Red-headed man preferred.—Miss. B., Muirhead, Alberta.
>
> Young lady wishes to correspond with cow-boy who can cook and keep house, lady musical and fond of travel. Photos exchanged. Ilene K. Box 3. Edmonton, Alberta.
>
> Handsome young Cowpuncher wishes to correspond with dark-haired young lady, one who can teach school preferred. Apply, High (Box H.) River. Alberta.

Who Will Take Me; -- ?
Crippled cowpuncher wishes old but nimble wife. Must be good cook and house-keeper as puncher has one leg off at knee, and a hook hand. Apply soon as possible to Robbers' Roost, Alberta.

Wanted—
Wealthy young wife, looks don't count. Man very homely but stylish dresser. J.D.B. Okotoks, A.

Wanted –
Wife. Must be good cook, able to fry steak and boil water. Not over thirty. Apply to C.H., High River.

Cowpuncher wants wife to run outfit for him. Has good house and a large set of unbreakable dishes. Box 322, Muirhead, Alta.

As the family ranches or ranch/farms rose out of the ashes of the cattle corporations and established the second cattle frontier during the first two or three decades of the twentieth century, their participants learned that collaboration was a necessary strategy for economic survival. This encouraged (or forced) them to understand that they were more resilient when working closely together in face of emergencies such as forest fires, just as they were when tackling routines such as putting up hay or nurturing and protecting their stock. Necessity, along with a numerical shortage of females, also influenced them in some ways to hold women, relatively speaking, in high respect while overlooking some traditional ideals.

This could and did at times have redeeming and gender-levelling effects. To quote an expert on the lives of western women during this period on the American frontier: "within the new and often unfamiliar sphere of activities imposed by frontier conditions, women compromised few of their Eastern-dictated goals, but they did find new outlets of expression and new fields for personal development and satisfaction."[37] Laura Macleay's case also provides support for this statement as well as Mary Kinnear's argument, noted above, that women in rural western

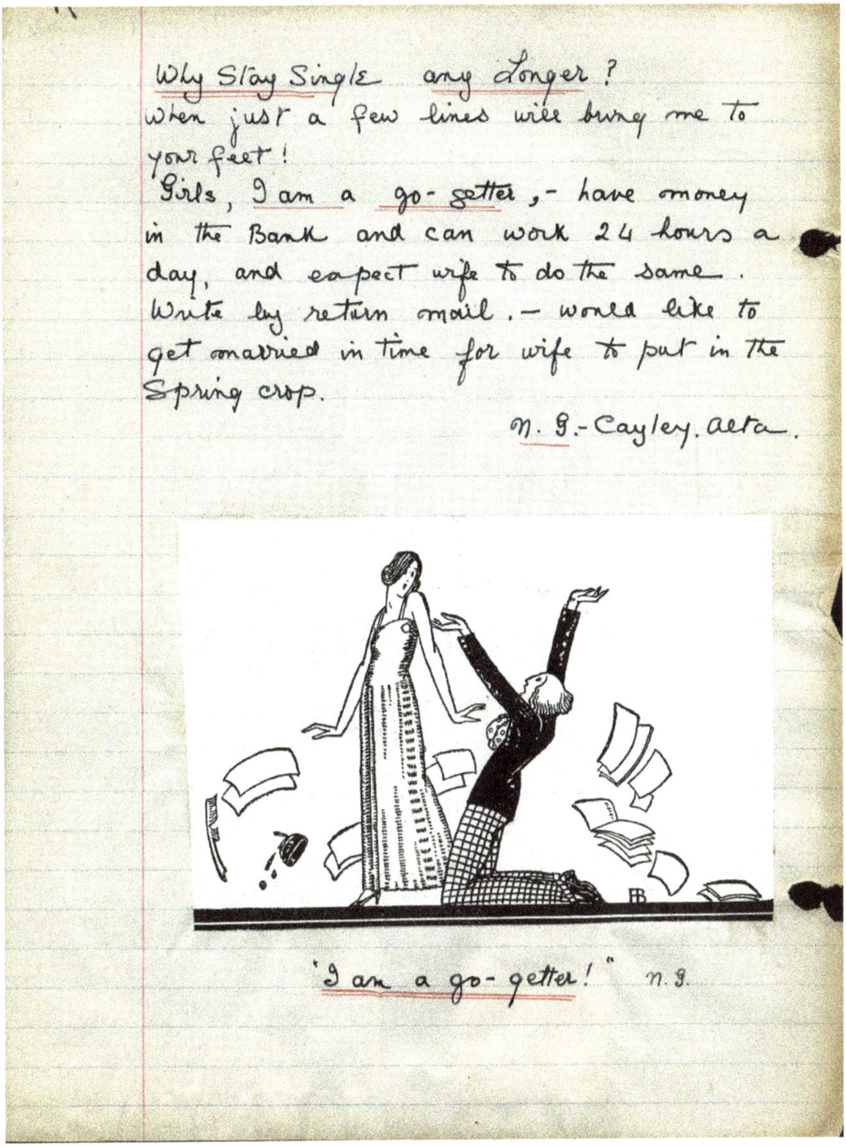
Why Stay Single any Longer?
When just a few lines will bring me to your feet!
Girls, I am a go-getter, - have money in the Bank and can work 24 hours a day, and expect wife to do the same.
Write by return mail, - would like to get married in time for wife to put in the Spring crop.

N. G. - Cayley. Alta.

"I am a go-getter!" N. G.

FIGURE 10.4. Humorous because it's not totally fictitious. "Matrimonial Bureau," *Rocking P Gazette,* March 1925, 72. Property of the Blades and Chattaway families and their descendants.

Canada during the interwar period gained a sense of their own stature on the land through their essential contributions to the rural economy.[38]

Again, this does not suggest that women on western ranches and farms achieved equality. The world was not ready for that. Even the men in the United Farmers of Alberta party, who did so much to promote women's rights including the vote in provincial and national elections and dower privileges, did not foresee the path to parity.[39] It does appear, however, that many rural women felt they were genuinely working in partnership with their husbands, which must, in turn, have brought some (perhaps many) of them a certain amount of confidence as well as self-esteem. This might also, then, help to explain why an almost endless list of western women the likes of Emily Murphy, Henrietta Edwards, Nellie McClung, Irene Parlby, Louise McKinney, Violet McNaughton, Ida McNeal, Mary McCallum, Georgina Binnie-Clark, and Hannah Gale were inspired to achieve so much politically, socially, or academically in the course of one lifetime.[40] It could, moreover, suggest a reason why women were able to win the franchise in the three prairie provinces before they attained it at the national level, despite the fact that the total population base (and especially the total female population base) of the three provinces was far smaller than that of Ontario alone.[41] Arguably, it was by living in frontier society and viewing what so many so-called ordinary women achieved in establishing and then sustaining the ranching and farming frontiers that these women were emboldened to believe in, and to act upon, their own vast potential. The fact that two very young ladies like Dorothy and Maxine Macleay had the temerity, albeit with their female teacher's help, to produce the seventeen monthly editions of the *Rocking P Gazette*, with all its art, current events, scholarship, wit, and humour, seems to bolster this interpretation.[42]

11

From Religion to Race

Just as the *Rocking P Gazette* helps to contest the myth of the male chauvinist cowboy in the Canadian foothills in the 1920s, it *ameliorates* the image of racial bigotry in rural Alberta as a whole. Prominent among the scholars who have created this image is Professor Howard Palmer, who has written of cases of discrimination against most non-Anglo groups in the province and in particular against Blacks and Chinese.[1] Other historians have also provided some very damning evidence of racism toward Indigenous communities: Hugh Dempsey vividly describes the most celebrated massacre—the infamous Cypress Hills affair of 1873—in which a number of American "wolfers," Canadian traders, and Metis shot up an Assiniboine camp, killing at least twenty men, women, and children;[2] in *Lost Harvests,* Sarah Carter shows that discrimination within the Department of Indian Affairs stood in the way of successful agricultural development on western reserves; further, in "Rangers, Mounties, and the Subjugation of Indigenous Peoples, 1870–1885," Andrew Graybill argues that the police do not deserve the praise some historians have heaped on them for bringing the reservation system to fruition, as he informs us that their prime objective was to cut the people off from their traditional food staple, the bison, ultimately subjecting the Cree and Blackfoot to a life of poverty and starvation;[3] and in *Clearing the Plains; Disease, Politics of Starvation and the Loss of Aboriginal Life,* James Daschuk, as his title suggests, takes that argument a step or two further.[4]

For the most part, people who advocated or constructed racial policies in the settlement of prairie western Canada were government or police officials in the East or they were residents of the bigger urban centres in the West, including Calgary, Edmonton, Lethbridge, and Medicine

Hat. They did not live in the countryside and did not experience the rural frontier first hand. What the *Rocking P Gazette* evinces is that the tempering (though certainly not the eradication) of racist attitudes, was, as with gender values, a reflection of the social environment in the countryside. One explanation for this we find in Paul Voisey's exhaustive study of the Vulcan area. Voisey shows that exclusivity by the English-speaking majority in that community was next to impossible in political, religious, social, and even sporting organizations because remoteness, the absence of particular amenities, and low population numbers (compared to urban centres) dictated that people of different racial backgrounds had no choice but to subject themselves to each other. This forced them to some extent to turn a blind eye to cultural idiosyncrasies.[5]

A recent book by Professor Abram de Swaan helps us see this from a new angle. De Swaan explains how, historically, bias against certain groups in various societies around the globe has at times been augmented through a process of "compartmentalization," even to the point of making genocide acceptable.[6] The author reveals that in numerous cases a minority group in a certain part of the world has been separated socially, politically, and psychologically by and from the rest of society and thus made to appear other than human. Using examples such as the German slaughter of Jews during the Nazi era and the massacre of Tutsis in Rwanda, he argues that segregation allowed people to be complicit or even to participate in mass murder, without pangs of conscience. In overly simplistic terms: it encouraged ordinary members of society to view the victimized group as a malicious, immoral, subhuman whole rather than a composite of individual men, women, and children. This enabled them to raise feelings of disdain to disgust, bias to extreme bigotry, and finally hatred to a justification for or acceptance of killing.

What happened in the Canadian West was just the opposite of compartmentalization. This is instructively illustrated in religious relationships. Out west, people of different Christian religions, cut off by crude transportation facilities from more densely populated urban centres and chronically short of capital, built and had to share very basic church facilities. This forced them at times to worship with Christian denominations other than the one they considered their own; and that encouraged them to see the people in those denominations as human beings, and thus to be somewhat accepting of them and less intolerant.[7] One example comes from the correspondence of Mary Inderwick, an Anglican who in the mid-1880s attended a Methodist sermon at Pincher Creek, Alberta, simply because it was the only one available on that particular Sunday. She disliked the service and especially the minister: "Though very earnest and suffering all sorts of hardships in his good cause, … he was the wrong sort of man to do much good to wicked humanity. He was learned in all but sympathy with young men's high spirits and love of fun and he lacked nice genial manners." When, however, an Englishman "attacked … the whole service" in a very "uncharitable way," she jumped to its support. "I glorified the whole church service including the awkward little man who preached to us." She acknowledged that in this she was empathizing with a Methodist friend who was at the church on that day.[8]

The *Rocking P Gazette* suggests that the Macleays themselves were among those who, in part because of the mixed-religion environment on the second frontier, were not particularly zealous in their commitment to a particular denomination.[9] To be sure, they were Christians. Among other things, in the December 1923 issue of the *Gazette*, a piece titled "Christmas" lays out the traditional story of the birth of Christ. "Nineteen hundred and twenty-three years ago, Joseph, the husband of Mary, had a dream" in which "he saw an angel from God standing beside him who said—'Joseph, I have come to tell you that you and Mary shall have a son sent by the Lord God. You shall call his name Jesus.'"[10] However, the lack of any but the most universal Christian expressions in the *Rocking P Gazette*, and the fact that in the several stories in the paper in which the church is mentioned it does not have a specific parochial label applied to it, seems to indicate that the Macleays, like Mary Inderwick, tended to be willing to look beyond denominational differences.[11] Family members did not let passionate commitment to one group or another cause strong feelings. They apparently were even prepared to play down the often-volatile Protestant–Catholic relationship in Britain. A line in an educational article in the *Gazette* glibly reads: "the people in the south of Ireland are mostly Roman Catholic and good fighters, while the people in the North are Protestants and quite peaceful."[12]

This sort of attitude in rural western society as a whole is visible in racial as well as religious relationships, and it derived from the same

FIGURE 11.1. Christian leanings. *Rocking P Gazette,* March 1925, 24. Property of the Blades and Chattaway families and their descendants.

source: subjection to diversity caused (or enabled) people to blur traditional boundaries. There is little doubt, for instance, that government officials and North West Mounted Police officers who actually themselves regularly patrolled the non-urban frontiers and spent much of their time with Indigenous people showed a willingness to protect them from white society and from the harsh policies they were supposed to be imposing. From the beginning, some police officers quickly extinguished any attempts by the cattlemen to exaggerate Indigenous depredations or to take the law into their own hands to deal with them. When George and Edward Maunsell and some of their neighbours blamed local tribes for the disappearance of their stock on range near Fort Macleod in the late 1870s, Indian Commissioner of the North-West Territories, Edgar Dewdney, backed by Colonel James F. Macleod of the North West Mounted Police, told them in no uncertain terms that much of this was undoubtedly due to cattle wandering back across the line to Montana and of rustling by whites. "Can we shoot any Indians we find killing our stock?" one of the ranchers asked. "If you do," Dewdney told them, "you'll probably hang."[13] In the 1880s, Colonel Macleod himself showed genuine compassion for the plight of tribal groups in southern Alberta who seemed to be on the verge of starvation after the disappearance of the Buffalo. He laid in all the provisions he could at Fort Macleod to help them out, and he pushed for more government generosity in other regions. He told his wife that he had "proposed that a supply [of foodstuffs] should be sent … to the Red Deer River to be given" the people there. The "poor creatures have been living on fish for some time back as the supplies here are about exhausted, but that source is now failing them as the fish are going back from the creeks to the lakes where the Indians cannot catch them with their nets." Macleod was highly critical of the government for not providing better. "I am not at all satisfied, with the arrangements [it] has made about the supplies for the Indians," he continued. "Indeed, they do not appear to appreciate in the least the calls that will be made upon them this summer. They appear to think that the poor creatures can gain their livelihood by hunting—as if everyone didn't know that there is nothing for them to hunt."[14] There is also evidence of police personnel reaching out to particular tribal groups and in so doing establishing amiable relationships with them.[15] Nanton area rancher Frederick Ings wrote of a Blackfoot Chief named Old Sun who, he believed, "became a loyal ally of the Red Coats," and actually called them "the Red Man's friends."[16]

Arguably, the moderation of bigoted attitudes toward people of a variety of racial backgrounds occurred among many of the cattle ranchers once they, like the Mounties, had been in the West and had experienced the rural second frontier setting for a significant period of time. The Macleays' example helps to illustrate this. There is no question that by the 1920s they were cognizant of the fact that the social environment they were living in was a multicultural one. In the western foothills, there was a strong numerical predominance of Anglos—English-speaking people from Great Britain, eastern Canada, and the United States. The Macleays were representative of those who had been "Canadianized." They were aware of their Scottish heritage, but after immigrating to the Eastern Townships in Quebec and then settling for decades in the West, they had gone through the process of acculturation. They must, therefore, at times have felt culturally distinct from even the more recent Anglo immigrants around them, whether from Great Britain, the United States, or even eastern Canada. And they could not have failed to recognize that there was a sprinkling of people in their society whose families had originated in a variety of non-Anglo countries of continental Europe.[17] Employed at one time on their own holdings were men with surnames such as Koff (Jewish), Krepps (German), Van Eden (Dutch), Weise (German), LeMaster (Americanized from the French, Lemestre), and Orvis (Gaelic).

The fact that Rod Macleay hired these people and that he collaborated with a variety of cultures both on and off his ranches suggests a certain imperviousness to racial differences. When he spent the day repairing a hayrack with an employee of German descent, or when he co-operated with a Dutch farmer to pasture his cattle in the wintertime, or perhaps build a fence between their respective landholdings, he could not help but recognize their personal characteristics and he lost any tendency to confuse them with some foggy vision he might have inculcated of an alien mass.

Macleay had a long history of working with Indigenous people as well as other non-Anglo groups. He knew of, and in his earliest years in the Alberta foothills probably paid into, funds that the stock associations provided to various bands to wage war against wolves that so severely threatened the ranchers' herds.[18] Thereafter, he also hired band members to work on his own place. A Stoney named Ezra Left-hand is mentioned in the September 1924 issue of the *Rocking P Gazette* as having stooked wheat in one of his fields.[19] Years later the *Lethbridge Herald* reported that Rod "employed … Indians" many times in "fencing, harvesting, horse-breaking and [for] numerous ranch duties." It is evident as well that he became empathetic toward these people when he got to know them. In the 1940s he and a number of his neighbours went out of their way to support the nearby Stoney band by putting pressure on the federal government to find them more land considered necessary to their survival. Finally, in 1948, Indian Affairs succumbed to their pressure by purchasing the Eden Valley ranch near Longview for the tribe. In gathering for an annual pow wow on their "traditional meeting place"—Macleay's Emerson homestead—the Stoneys "spoke in gratitude … to the white people who had aided them ably in representation to government." The *Lethbridge Herald* asserted that "no private citizen has so consistently befriended the Stoneys and supported their claims," as Macleay. The "Indians" apparently emphasized that without [Macleay's] "help they could not have got through some of the hard winters."[20]

Laura Macleay no doubt adjusted somewhat to cultural realities on the frontier too, via her own subjection to diversity, and, perhaps to some degree, in the process of assimilating her husband's views. The same could be said about Dorothy and Maxine, who, as we have seen, worked cattle with various of the Macleays' cowpunchers as they grew up. It is evident, moreover, that all three members of the *Rocking P Gazette's* staff had reasons directly associated with the publishing business for embracing a multicultural attitude. As with any frontier newspaper, in order to achieve credibility, they needed the collective support of their local community. This they instinctively understood, and they did their best to keep everyone on the ranch involved in the paper on a continuing basis by, for instance, encouraging them to enter little monthly contests to pick a title for a story,[21] or to name a fictitious person,[22] or simply to pick a number from 0 to 99,[23] or count dots.[24]

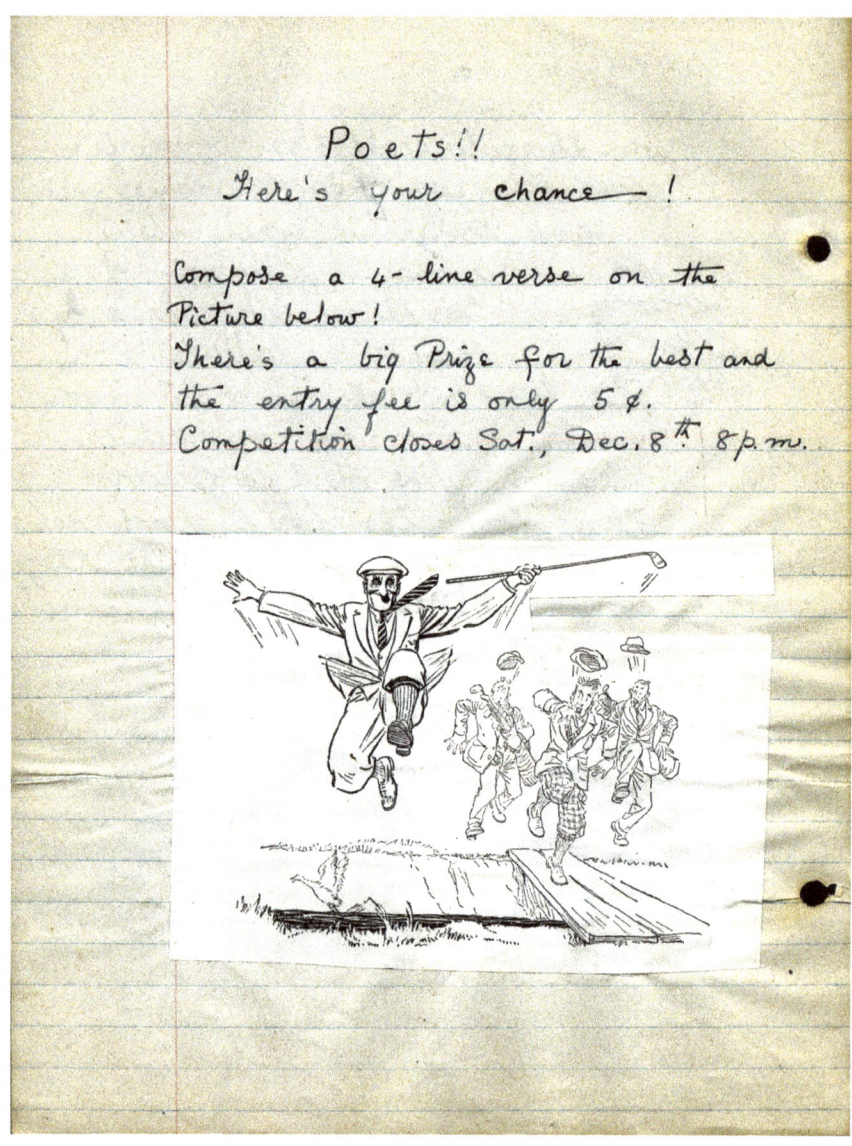

FIGURE 11.2. One of the girls' little competitions to keep the men interested. *Rocking P Gazette,* November 1923, 54. Property of the Blades and Chattaway families and their descendants.

The fact that the ranch hands contributed so many of their own written pieces to the publication indicates that they found it engaging. However, at points when their work schedules were intense, their contributions to the *Gazette* tended to decline, as they did to the Foothhills Terrors. At one such time, Dorothy and Maxine softly and indirectly reprimanded them for not showing more enthusiasm.

> <u>Last month</u> the overworked staff of the *Gazette* burned the midnight oil (two whole whales full) in endeavour to meet the pressing demands of a public hungry for samples of our best literary food.... We accomplished <u>something</u> but there were no outside contributions and no solutions to our contest. <u>This month behold</u>!!!!! Owing to the energetic agency at the Bar S Ranch of our cowboy reporter, <u>Frank Van Eden,</u> and the enthusiasm which he has inspired in all departments, punchers, farmers, cooks and bottle-washers—we now place in the hands of our admiring circulation a champion number. N.B. New poets, New Artists, Record competition results, More Local News....[25]

In essence, the *Gazette* was a microcosm of its society. Like that society, it depended on co-operation and collaboration between the owners (Dorothy and Maxine), the manager (Miss Watts), all the men ("punchers, farmers, cooks and bottle-washers"), and anyone else who was part of the community around it (i.e. on Macleay ranches). To be successful it too needed the dedication and commitment of the Van Edens, LeMasters, and Kreppses as much as those of the McDonals, McKinnons, and Smiths. It is not really surprising, therefore, that the *Gazette* tended to be a voice for racial harmony. The reader finds all kinds of evidence of this. For instance, the jokes about nationality aimed at the numerically predominant culture on the ranch—the one the paper's founders themselves claimed—are numerous. The editors' pen poked fun at Scottish people, not just, as we have seen, for their language but also:

For their supposed parsimoniousness:

> "A Scotchman, wishing to commit suicide, went next door and carefully turned on his neighbor's gas."[26]

And sometimes more generally:

> "An Englishman, a Scotchman and an Irishman had a wager concerning which of them could stay longest in a room with a skunk.
>
> The Irishman went in first, stayed one minute and beat a hasty retreat.
>
> Then the Englishman, with his superior powers of endurance, remained in the room for two minutes before he was beaten.
>
> The Scot won the wager, as he stayed in for three minutes—then the skunk came out!!"[27]

A racially broad view is also reflected in plots and story lines the girls and their teacher presented. Above all else, these show genuine affection and concern for Indigenous people. An excellent example is a piece by Watts in the November 1924 issue in which she made ranch worker, Ezra Left-hand, the hero.[28] In describing Left-hand's marriage ceremony, Watts wrote:

> [It] was held under the blue Skies, among the flaming glories of the Indian Summer and to the glorious name of Left-hand the preacher added that of "Ezra" which means "Help"—For he prophesied that in years to come ... [he] of the strong Left Hand [would] do much to help his brothers, the Redmen, in the days when whites should ride and herd cattle in the Foothills of the West.[29]

This same recognition of the peril white folk wrought for Indigenous people and their culture as they invaded their territory is evident in the "Song of Hiawatha" credited to Annabella Trunk in the May 1924 issue.

FIGURE 11.3. Depiction of a Scottish gentleman. *Rocking P Gazette*, November 1924, 34. Property of the Blades and Chattaway families and their descendants.

It draws on Henry Wadsworth Longfellow's 1855 poem of the same name. "But with spring came food, and people from the far east who told Hiawatha that the white man had come. That very day across the great lake came the 'canoe with pinions' full of white people. Then Hiawatha bid good-bye to old Nokomis, and got into his birch bark canoe and sailed away into the west alone to die."[30]

Dorothy, Maxine, and their teacher depicted Indigenous people as heroic worthy of compassion and understanding. The Macleay girls' visual art adds to this picture. It looks back to a perceived age when they were warriors—gallant, forceful, self-determined—or hunters and traders—freely and purposely participating in the fur trade.

The Macleays' attitude toward the Chinese workers they employed in the 1920s, as reflected in the *Rocking P Gazette*, actually suggests a level of racial compassion uncommon almost anywhere in the West at the time. It also helps to illustrate the thesis promulgated in this chapter. In order to understand this it is necessary to recognize that in tiny prairie towns as well as the bigger urban centres, the Chinese were the one people against whom prejudices were almost never overcome.[31] They were constantly restricted to the bottom rung in the social hierarchy. White people were prepared to use their laundry services or eat in their restaurants, but they did not want the Chinese to mix and mingle socially or politically with the rest of society. Some Albertans went so far as to rally publicly against allowing white women and girls to work in Chinese business establishments, presumably for fear the owners would infiltrate white blood lines.[32]

The *Gazette*, though, reflects a genuine warmth for the two Chinese men who cooked the meals in the Macleay home during the early to mid-1920s. Among other things, it features a piece in the February 1924 issue written in Mandarin by Charlie Lung, with a painting of a Chinese junk next to it.[33] Below the painting are Rudyard Kipling's poignant and race-transcending words: "Till the junk-sails lift through the homeless drift and the East and the West are one."

CHINESE DAFFODILS ARE THE MOST BEAUTIFUL OF VISTAS

I have heard it said, "The people are the basis of the state, and food is [as important as] Heaven to the people."[34] [I add that] Food among the people cannot be lacking for [even] one day.

FIGURE 11.4. Dorothy Macleay's depiction of an Indigenous warrior. *Rocking P Gazette*, April 1925, cover. Property of the Blades and Chattaway families and their descendants.

Figure 11.5. Indigenous riders. *Rocking P Gazette*, February 1925, cover. Property of the Blades and Chattaway families and their descendants.

Figure 11.6. Indigenous history as Dorothy and Maxine knew it. *Rocking P Gazette*, April 1924, 30. Property of the Blades and Chattaway families and their descendants.

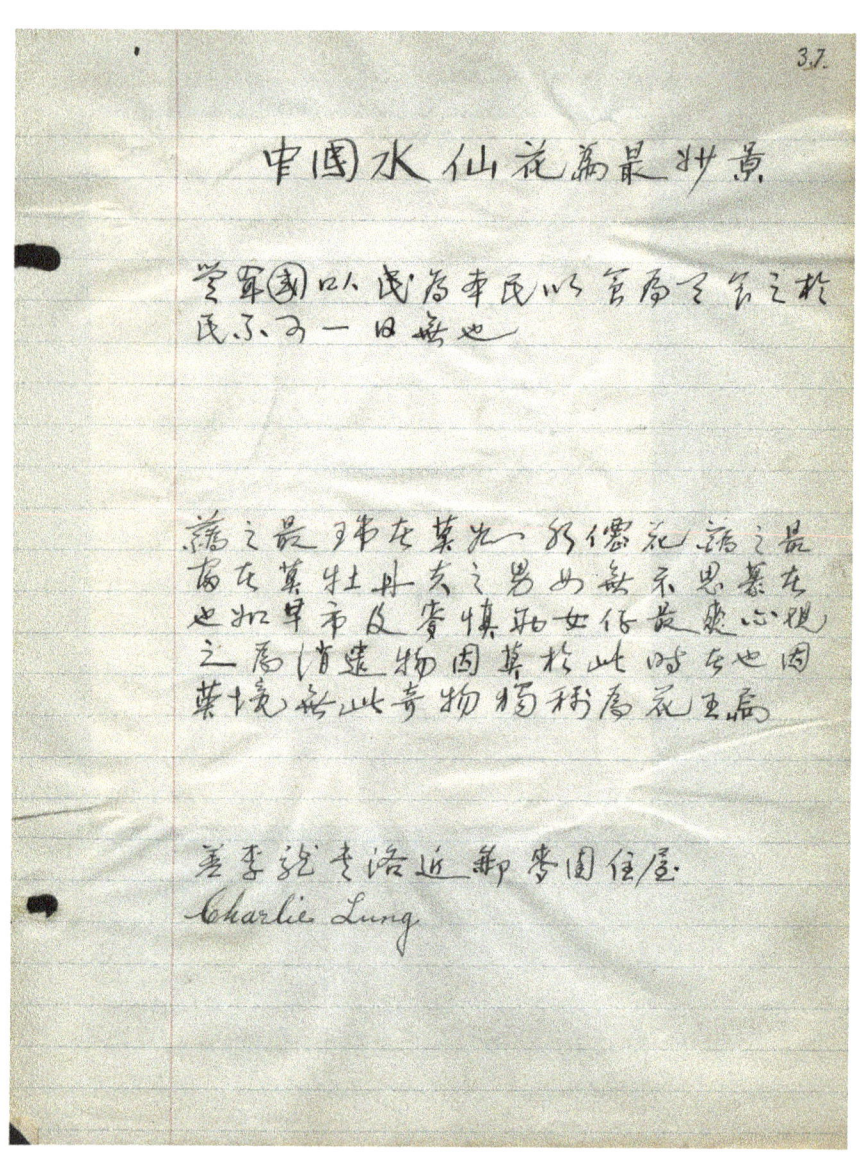

Figure 11.7. Charlie Lung's writings. *Rocking P Gazette*, February 1924, 37. Property of the Blades and Chattaway families and their descendants.

Of the most auspicious of flowers, none can compare with [Chinese] daffodils; of the most luxuriant of flowers, none can compare with peonies. Among the sons or daughters of men,[35] there are none who do not pine and long [for Chinese daffodils]. In going to the morning market and in arriving at Maishen,[36] sick and frail girls are most refreshed and pleased and see [Chinese daffodils] as objects of diversionary amusement, because no time is better than this[37] [to enjoy the beauty of Chinese daffodils]. Because the realm of the British does not have these marvelous things,[38] they alone are called the king of flowers.

[Written by] Charlie Lung at Maiwei,[39] on the outskirts of [characters illegible].[40]

The Macleay girls also included Charlie Lung in the little monthly competitions they ran in their newspaper and, most remarkably, their teacher praised Lung and another Chinese cook, George Wong, in the short story "Forty Years On." In this story, Watts described the family and only a few of its closest workers coming back for a reunion at the ranch in 1964.[41] "Supper was served on the hillside and the dear old cooks, Charlie and Wong, surpassed themselves in the excellence of the repast. Wong, the Chinese cow-boy-cook, had for long years been famous in the wild and woolly West. Charlie smiling as ever, called Hullo! To all his old friends, while he served out hot cakes and 'lasins pie.'"

Unlike most people in western society, the Macleay girls and their teacher shared a home with their Chinese acquaintances and got to know them on a personal level. To borrow from both Voisey and de Swaan, in this situation exclusion and compartmentalization were impossible, and long-standing racial sensibilities somewhat irrelevant. The objective here is not, and has not been, to paint the Macleays, or any westerners, in eulogistic terms. We are not suggesting that anyone purged themselves of Old World racial prejudices, only that in the frontier setting some people found it necessary to temper them. To put this another way: men like Colonel Macleod may have felt compassion for Indigenous people, but

FIGURE 11.8. Ancient Chinese sailing ship or "junk." *Rocking P Gazette*, February 1924, 36. Property of the Blades and Chattaway families and their descendants.

FIGURE 11.9. Charlie Wong and other Macleay workers as the *Gazette* staff thought they would look in forty years, featured in the same issue as Watts' story, *Rocking P Gazette*, May 1924, 77. Property of the Blades and Chattaway families and their descendants.

that did not mean that they were able, or even tried, to think of them as equal to themselves. "I have the room formerly occupied by Captain Walsh [at the Cypress barracks] and sleep in his bed," Macleod wrote to his wife in 1878.[42] "I cannot help thinking of the queer companions who must have occupied it with him. I turn from the thought with a shudder. The idea of a dirty squaw in the place of the sacred person of one's wife. Isn't it beastly—that's the word that suits the case exactly."[43] We can assume that the Macleays retained biases toward other races as well. It was, moreover, the condition of their social and economic environment, not some innate, forward-looking, enlightened view of the world, that forced many people to soften some of their ideas on race or, indeed, gender. Amelioration of Old World values was a necessity, or at very least helpful, in a remote and as yet relatively undeveloped agricultural society where people routinely had to transcend traditional boundaries in order to confront challenges such as putting out a prairie fire, feeding a number of hungry cowpunchers, or putting together a "Conservatoire Musicale."

Frontier historians have arrived at this conclusion when viewing other societies during the building years. For instance, scholar and historian of American frontier women's history Sandra Meyres writes: "many of the irrational myths and stereotypes which formed an important part of racist and nativist sentiment in the East and South were also present in the [American] West." However, "the face to face confrontations and political, religious and economic conflicts—which were also an important part of nativism did not occur as often in frontier areas or in the new Western communities." "Westering women," Meyres concludes, "packed many of their prejudicial myths and stereotypes with them, but often these ideas were modified or sometimes discarded, along with other prized possessions, when they confronted the frontier and its inhabitants" first-hand.[44] In conjunction with other source materials, the *Rocking P Gazette* enables (or requires) the historian to depict the Macleays and their life on the ranch as they really were, rather than as we might assume them to have been. That is one of its most substantive qualities.

It would be convenient to end the discussion of race in frontier society and in the *Rocking P Gazette* at this point. Unfortunately, that is not possible. In dealing with the Macleays, their society and attitudes toward ethnicity, we cannot avoid considering evidence in the *Gazette* that on the surface detracts from our argument. That is to say, there is some discomfiting language that if taken in the modern context could, and would, be considered extremely racist. To account for this and maintain our argument, we need first to remind the reader (and ourselves) that present-day values may differ from those of earlier times. When she, he, or we come across words that now might be viewed as racially derogatory, it is necessary to recognize that the people who used them in the 1920s did not necessarily intend them that way. There are two examples: in suggesting the cause of a forest fire, a report in the April 1925 issue of the *Gazette* noted the possibility that it occurred when "an Indian Squaw's teepee was blown over in the gale."[45] In the nineteenth and early twentieth centuries, white people often used this word. When it was not accompanied by negative adjectives, they simply meant to signify an Indigenous woman.[46] Indeed, to cowpunchers in particular the word was often a term of endearment. In Montana, Teddy (Blue) Abbott became infatuated with an Indigenous girl who would have nothing to do with him.[47] "I wanted this girl so much that I asked her if she'd marry me, but she wouldn't do that either." He explained that he and his buddies "were starving for the sight of a woman, and some of the young squaws were awful good-looking, with their fringed dress of soft deer or antelope skin that hung just below their knees. ... Oh, boy, but they looked good to us. But I was always that way. I always wanted a dark-eyed woman."[48] Abbott eventually married the daughter of a white rancher named Granville Stuart and his full-blooded Shoshone wife. By all appearances, they had a very happy marriage—producing eight children and numerous grandchildren.[49]

Abbott's matrimonial harmony is actually replicated in a story in the *Gazette* titled "In the Tall Timber," in which a young man named Hal Brown falls in love with and marries a "beautiful" "half-breed" named "Melesse," who stays with him through difficult times on the run from the police, and nurses him for weeks through a dreadful disease, the "Red Fever."[50]

The famed cowboy artist, Charlie Russell, felt deep and genuine admiration for the Indigenous men and women he met in the American and Canadian West. He painted the women many times, and he used

Figure 11.10. Charles M. Russell, *Squaw Travois*, 1895, watercolor, Montana Historical Society Collection, X1954.01.01, Gift of Maude and Florence Fortune.

the now-offending "S-word" with anything but contempt or disdain. To quote Russell's best-known biographer, he treated the women "with poignant tenderness" in his paintings. "In Indian Maid in Stockade," he "brought a unique interpretation to a portrait of a tribal beauty. The woman slouches against the silver-grey logs of the fort, flaunting not only her handsomeness, but her ornamentation as well. Russell imbued her with independence, self-assurance, and almost a touch of defiance, making the viewer aware of the woman's sense of self-worth."[51] Charlie Russell, like Teddy (Blue) Abbott and all the Macleays, lived and sometimes worked closely with Indigenous people. Once again, we would not suggest that he or they had no biases toward them but simply that they could not help but see and feel that they were people too.

Another word the Macleays and many others commonly used that would today be considered not just racist but extremely so was the "N word." It would also be wrong, however, to simply label this as such in either the first or second cattle ranching frontiers. The word was regularly affixed to John Ware and his ranching outfit, but we should remember that Ware himself was one of the most widely respected and time-honoured men in the western Canadian ranching community, mainly because he put his celebrated rangeland skills to work establishing the cattle industry.[52] "John is not only one of the best natured and most obliging fellows in the country, but he is one of the shrewdest cow men, and the man is considered pretty lucky who has him to look after his interest," reported the *Macleod Gazette* on 23 June 1885. On 2 March 1892, the *Calgary Tribune* stated that "no man in the district has a greater number of warm-personal friends."[53] Slim Marsden of Vulcan once honoured him as "a famous Bar U cowboy, who in his heyday rode the wide open spaces, broke uncountable wild broncos, started two ranches of his own," and was "known where ever stockmen gathered together, as a top hand, the great and colourful personality—John Ware."[54]

The cynic would say that Ware was accepted because, as was so often said, though he was "black on the outside he was white on the inside."[55] But in a way, that is just the point. Before becoming a family rancher, he had co-operated closely with white cattlemen, first in driving in stock for the Bar U way back in 1882 and then as a cowhand working some of the first great herds on the open range. He was known to be gifted at all the specialties of the craft he and the other cowboys shared—"the best specimen of the negro race who came as an expert cowman and bronco-twister ... a splendid roper and a grand roughrider."[56] The men who worked alongside Ware were forced to see him as one of them. They therefore adjusted their attitudes toward black people—a little. Were all black people, like Ware, equal to whites in at least some respects? No—but obviously, the possibility existed that some could be. Racism did not go away—it was tempered.

The tempering rather than the obliterating of racial prejudice toward black people is also demonstrated by some of the humour in the *Gazette*. Thus under "Jokes" in the September 1924 issue the following dialogue appears:[57] "When the colored couple were being married, and the minister read the words, 'love, honour and obey,' the bridegroom interrupted, 'Read dat ag'in; parson; read dat onst mo' so de lady kin ketch de full solemnity of de meanin. I'se been married befo.'" This is just one of a number of times the editors put such language in the mouth of a black person.[58] Unquestionably, it illustrates the acceptance of stereotyping based on unconfirmed and, therefore, prejudicial assumptions. However, it does not mean that the girls or their teacher hated black people. It just means they found what they assumed was the typical dialect of that particular racial group amusing. Exactly the same could be said about their rendition of the dialect typifying their own and a number of their workers' countrymen.[59]

First Scot – "Wot sort o'minister hae ye gotten, Geordie?"

Second Scot – "Oh well, he's muckleworh [worthless]. We seldom get a glint o' him; six days o' th' week he's envees'ble and o the' seventh he's incomprehens'ble."[60]

The *Gazette* illustrates too that the N word was sometimes (perhaps often) used as, or in, the name of a black horse. The obituary for a cowpuncher in the April 1925 issue reads: "For a number of years recently Bill Krepps has been on the Rocking P continually, except for a few months in each year spent in Calgary, during the coldest winter months. He will be best remembered, riding 'Old Nigger Baby' down the trail at a slow walk, with little Buster bustling along at the cowhorse's heels"[61] When we judge this, we should acknowledge two facts. First, there was almost nothing in this world a cowpuncher cared about more than his horse—often his pal as well as his trusted mount—the one being that could help make him great in a world of riding, roping, and cutting. His or her name was a form of endearment, not an insult or fumbling effort to equate a certain type of people with animals. Secondly, people in the western foothills in the 1920s were isolated by time, great distances, and bad roads from the racially fuelled world of the southern United States. When Anglos used this expression, it reflected their naivety, their ignorance—but it did not necessarily signify aversion to, or loathing of, people whose predecessors came from Africa. In essence it just meant black.

Times change, social conditions change, meanings change. When the N or the S word is used today it is undeniably, blatantly, and brazenly racist. In the remote and isolated rural West of the past, when all sorts of people from all sorts of places had to moderate Old World presumptions for practical reasons, this was not necessarily so. Our next chapter will underscore this argument by examining and trying to explain the *Rocking P Gazette*'s vision of self.

12

Reinforcing Family Values

It is seldom appreciated how secluded country life in western Canada was until after World War II. Though the automobile and the trucks for hauling grain were common by the 1920s, they were capable of doing only a little to firm up connections between rural and urban communities or, indeed, between rural and rural communities. Firstly, they could not make the roads any better. Clay Chattaway rightly describes them as "trails" rather than what we know today as roads—they were muddy and slippery when it rained or during spring thaw; they were also not raised above the fields and pastures around them as they are today, and, therefore, drifted over badly during even relatively modest winter blizzards. They continued to be crisscrossed with livestock gates too, which someone in every car, truck, or wagon had to open before going through and dared not leave that way afterward. Moreover, the motor vehicle in the 1920s was still a very basic piece of equipment. In 1924, the Gazette reported that Rod's cousin had bought a car. He was none too pleased with it. "A great sensation was caused by Stewart Riddle, one day not long ago, when he drove into the Rocking P, in a tin fliver [sic] and claimed it was his own, and that he only paid $20.00 for it.[1] Later on he said if he had had an axe he would have killed his car coming up some of the hills."[2] To be sure, rancher/farmers felt that the first trucks were an improvement over horses for hauling grain, but the vehicles nonetheless were not comparable to the much bigger and more powerful modern versions running on open gravel and asphalt surfaces.

Relative rural seclusion is evidenced by the frequent references in the *Gazette* to the T. Eaton and Simpson's department store catalogues for mail ordering consumer goods.[3] Rural people ordered all sorts of things

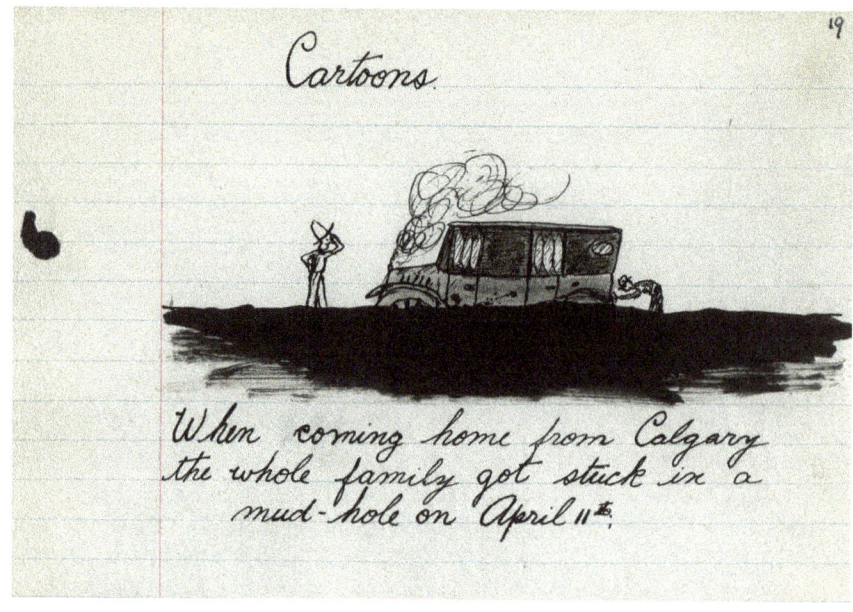

FIGURE 12.1. Ranchers liked the trucks in the 1920s, yet they continued to haul wheat by horse and wagon when the roads and paths were just too bad for motor vehicles. *Rocking P Gazette*, Part A: April 1925, 19; Part B: November 1924, 24; Part C: September 1924, 11. Property of the Blades and Chattaway families and their descendants.

through the catalogues rather than enduring a long and cumbersome trip to one of the big towns (in this case Calgary) every time they wanted to search for products. The *Gazette* thus ran mock ads for everything from "Electric pads ... for punchers' backs when cold," to overcoats, to engagement rings, to "dancing pumps," and "Miss Lovely writing pads."[4] Purchasing through the catalogues was a little like using the online system today except that one had to rely on the chronically slow regular mail service and eventually probably had to do some travelling anyway to pick up the items at a depot in the nearest town or village.[5]

Part of the attraction of the catalogue, particularly for women, was the sensation they got, while thumbing through its pages, of being able to keep up to date on all the latest fashions in the big cities. Ranch women may have been proud of the country life, but they did not want to be considered, or to consider themselves, "country bumpkins." One woman remembered that it was in part through the ads in catalogues and other print media forms that she managed to enter the world of "shops and fashions and coloured illustrations of the Gibson Girls with high pompadours and shirtwaist dresses." Eventually, she said, "Tiffany's, Fifth Avenue, Delmonico's, the brown stoned mansions of the Vanderbilt's and their ilk" in faraway New York, "seemed as real to me as Nanton."[6] Dorothy and Maxine shared this fascination, and they fed it by replicating images from the catalogues and presumably from any magazines to which they had access on the ranch, for the benefit of their female audience, including Miss Watts, their mother, Aunt Gertrude, who so often stayed at the ranch after her husband died (and then married Stewart Riddle), and any of the females, including Mrs. Walters and her daughter and other wives who lived with their husbands on the ranch as well.[7]

The Macleay editors' utilization of the catalogues and the mail order system to overcome the remoteness of country life fitted in with their positive approach to the racial and cultural diversity that existed on the ranch. It is also evident that they hoped to enhance worker contentment and thereby facilitate productivity on the ranch as a whole. Rod Macleay might well have realized that one of the advantages his enterprise had over the failed company operations was the close working relationship he was able to cultivate not just with Laura and the girls but with all his employees. Whatever the case, as we have seen, he worked closely

FIGURE 12.2. Latest fashions in hair styles were also of interest to the female audience probably taken from magazines. *Rocking P Gazette*, Part A: February 1925, 65; Part B: February 1925, 65–66; Part C: March 1925, 68. Property of the Blades and Chattaway families and their descendants.

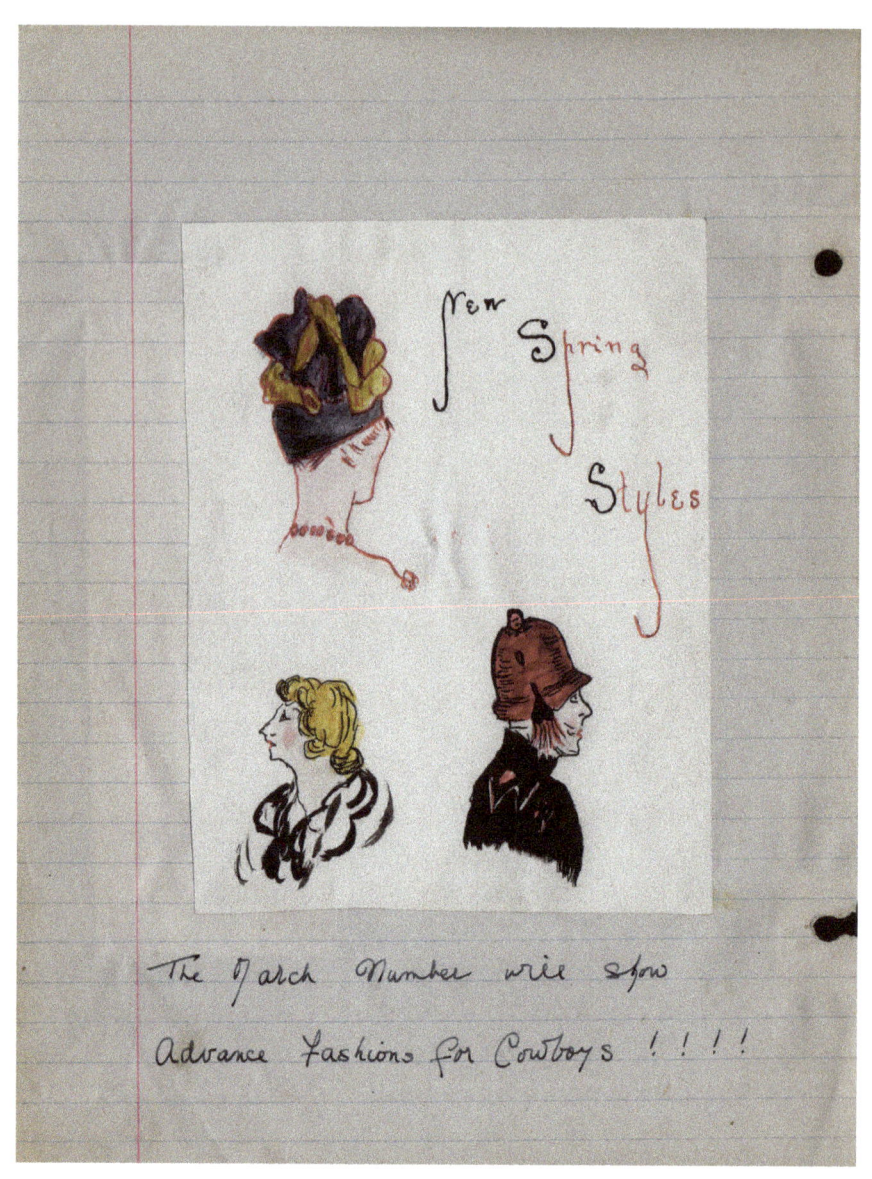

The March Number will show Advance Fashions for Cowboys !!!!

Ads.

The latest in hair-cuts for the fashionable women.

The best and "handsomest" bit ever made. Claude Mills. High River.

Poets!! Look at this, a rhyming dictionary now all your troubles are over. Apply Rocking P Gazette staff.

Read this now.

Cowboys there will appear, only one more Gazette, so come "across" with something to fill "er up. We want a hundred pages in the April number.

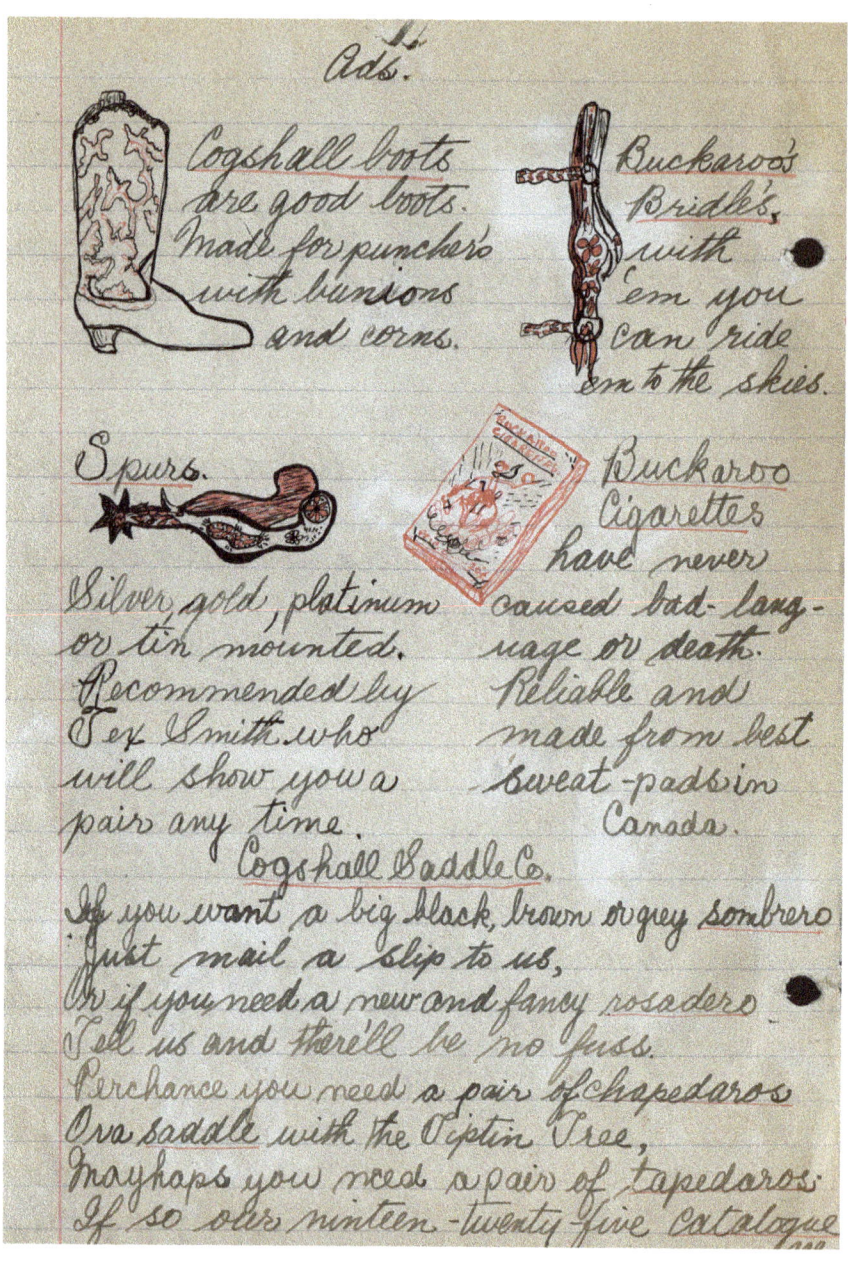

FIGURE 12.3. There are catalogue ads in the Gazette for men's things too—but fewer. "Ads," *Rocking P Gazette*, February 1925, 68. Property of the Blades and Chattaway families and their descendants.

with them often and whenever he could. However, the Macleays' holdings were so large, relatively speaking, and the number of employees so great that Rod could obviously not be in close contact with all his men on all parts of the operation all of the time. He had, moreover, obligations that periodically took him off the ranch, for instance, to deal with the bank or the courts or to fulfill his obligations as vice president of the Western Stock Growers Association or as president of the Council of Western Beef Producers.[8] Anything that would help to build up incentive and a sense of duty and commitment among the men when he could not be there was important. As one reads through the *Rocking P Gazette* one feels quite certain that under their teacher's guidance, Dorothy and Maxine recognized this and that by drawing on the contributions of all the young "punchers, farmers, cooks and bottle-washers" they ultimately were trying to bolster and sustain the collective feeling among the Macleay workers that they were far more than just a wage-earning necessity—they too were part of a family.

This is demonstrated in a number of ways in the pages of the newspaper. Firstly, there is a "Local News" section in each issue, which duplicated a section that appeared in regular small-town newspapers in this period, variously titled "Of Local Interest,"[9] "Local and Personal,"[10] "The Home Town,"[11] or "Local Notes."[12] Under such titles the community papers reported any bits of current events, however trivial, that acknowledged almost anything in which individuals in the area were involved. The following sample from the February 1924 *Gazette* reads very much like any of those reports:

> Feb. 15th. Miss E.B. Watts while out sliding with D. Macleay, met an accident. She was thrown from the toboggan, and seriously damaged her knee cap. However she brightened up as [her then boyfriend] McKinnon hove into sight, *over the horizon.
>
> Frank Van Eden rode over from the Bar S on his new horse "Spoof," on the 15th.

Jesse Walters and Ted Nelson[13] came over to "fan" wheat on the 23rd. They were unable to do any work that day, but got one load done on the 24th. On the 25th they had engine trouble and it took S. Riddle, J. Walters and T. Nelson all day to "get'er going" again.

Feb. 26th. Robert Raynor and Stewart Riddle went over to the Bar S to chop [grain]. They worked all day and finished up their job on the morning of the 27th.

Feb. 28th, Val Blake and Tex Smith rode the Bar U flat, bringing home a few strays.

Stewart Riddle this month won a thirteen dollar "Cayuse Indian Blanket," by picking out the right name in a competition given by the Shriners in Calgary.[14]

A significant proportion of this type of information came from one or two of the workers appointed by the Macleay girls, and that, along with the news itself, could only have created and augmented a feeling of commonality—of belonging.[15]

The *Gazette* staff no doubt recognized, too, that portraying an individual Macleay ranch hand in any of the pieces they themselves wrote added to the effect. For that reason, many of their short stories centre around Tom McKinnon, Jimmy Hendrie, Val Blake, Tex Smith, Bill Krepps, Frank Van Eden, Stewart Riddle, or Jim or Ralph McDonal. Usually, one of them is featured in an amusing vein, as a cowboy in love or a vicious gunslinger or just a young ranch hand prone to foibles of one type or another.[16] Sometimes, on the other hand, as in Watts' "Forty Years On" and "The Romance of Ezra Left-hand," one or more of them is shown in a more serious aspect.

The ladies liked as well to employ humour as a tactic to keep the people in their audience laughing lightheartedly at each other. "Crack-shot Val," credited to "Dynamite Dick" but in Maxine's hand, takes a shot, so to speak, at Val Blake for an incident about which he was no doubt feeling a bit sheepish.

The coyotes were not very thick,
So the hunters never got a pick.
And there was a lot of fun
When anybody sighted one.
Now hold your breath and listen well,

To the tale I have to tell
'Twas a bright, sunshiny day,
Clem and Val, were hauling hay.
Val was feeling extra strong,
He had his faithful gun along.

All of a sudden they saw on the hill,
A Coyote, who stood very still.
"I'll shoot that bird, hold the horses tight."
Val grabbed his gun and took an aim, right
At the head of the coyote, so staunch
Who was now sitting on his haunch.

Three dreadful minutes silence reigned,
And then the gun roaring flamed,
And kicked Val out upon the ground.
While Mr. Coyote got up and looked around.
Then slowly trotted off on his way,
As limping Val climbed back on the load of hay.[17]

In "Val's Lament" by "Sixshooter Sam," Dorothy teased a cowpuncher about his love life:

Aw, Minnie, Minnie, where art thou?
Thou whom I think of when driving the cow,
Thou whom I dream of when working the plow,
And many moons have passed
Since I saw you last, Minnie.

> Minnie, who used to laugh and talk,
> Minnie, who pointed to me the love hawk,
> And darned, with her tender hands my sock!
> But many moons have passed
> Since I saw you last, Minnie.
>
> Come, and once more, let me behold
> Your shining face of brilliant gold,
> Cause Minnie my hand is growing cold,
> But many moons have passed
> Since I saw you last, Minnie. ...[18]

Dorothy and Maxine also used images to poke fun at individual Macleay workers. The following and others are found under "Cartoons," in the April 1925 issue (see page 197, 198, and 199).

One of the clearest single expressions of the *Rocking P Gazette*'s team-building qualities is a limerick entitled "How d'ye do" by the "Classic Muse" (Ethel Watts), which was based "after the latest musical hit of the city" (perhaps *Alice in Wonderland*). All family members and all the ranch hands are given credit in the poem for their place in the Macleay community and in a western vision. Even the family pets are included. The poem is a fitting conclusion to this part of our discussion.

> How d'ye do, rocking P?
> How are you?
> How's the cattle,—how's the wheat and oat crops too?
> Is the Home Place blooming still?
> How's the Calf Camp cross the hill?
> How d'ye do, Rocking P?
> How d'ye doodle-doodle-do?
>
> How d'ye do, Rocking P?
> How are you?
> How d'ye do, Boss Macleay?
> How are you?
> How's the prospects of the market look to you?
> Have you made your pile in beef?

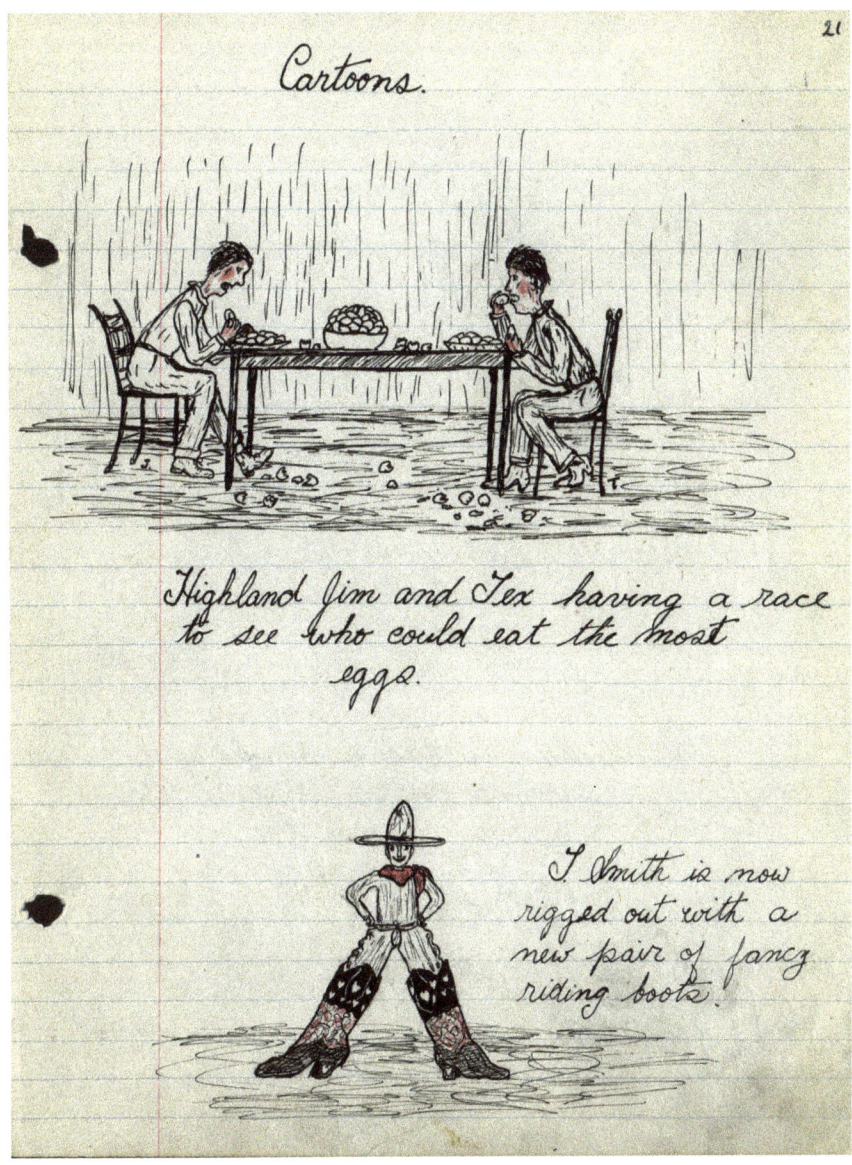

FIGURE 12.4. "Cartoons," playing on cowhand foibles. *Rocking P Gazette*, Part A: April 1925, 24; Part B: April 1925, 42; Part C: April 1925, 25. Property of the Blades and Chattaway families and their descendants.

Is you wealth[y] beyond belief?
How d'ye do, Boss Macleay?
How are you?

How d'ye do, Madame Macleay?
How are you?
Have you had your longed-for trip to Honolulu?
Do you still devour the Red Book?
Have you found the perfect cook?
How d'ye do, Madame Macleay?
How d'ye do?

How d'ye do Mrs. Doc?
How are you?
Are you fas-cinating still to each Bob and Bill Hugh?
Do you ply your needle still.
With grace and dainty skill?
How d'ye do, Mrs. Doc?
How are you?

How d'ye do Stewart Riddle?
How are you?
Do you still love macaroni and oyster stew?
Do you still enjoy John Haig [scotch whiskies]?
Have you joined the Temperance League?
How d'ye do, Stewart Riddle?
How are you?

How d'ye do, Dorothy?
How are you?
Do you ride the wiley broncs, cayuse and hot-blood too?
O'er the hills do you still lope?
With your chaps and spurs and rope?
How d'ye do, Dorothy?
How are you?

> How d'ye do, little Maxine
> How are you?
> How's Sancho—how're the cats and turkeys too?
> Does poor "Sawndy" linger on?
> Have a heart—let him be gone!
> How d'ye do, little Maxine?
> How are you?[19]
>
> How d'ye do, all you Boys?
> How are you?
> Tex, Val, Jimmie—Bill and the rest of you!
> Here's the very best of luck!
> Let'er rip and let er buck!
> Here's to you, all you boys!
> How are you?
>
> How d'ye do, great wide West?
> How are you?
> How's the foothill range 'neath skies of radiant blue?
> Here's to all who love the West!
> Where life goes with a zest!
> How d'ye do, great wide West?
> How d'ye doodle-doodle do?
> How d'ye do, great wide West?
> How are you![20]

It is instructive that the above poetic piece, expressing such an emotional connection between the Macleay family members and a number of the people who worked for them over years, was written by one of the people who worked for them for a considerable period of time. Precisely how successful the Macleays were in retaining the loyalty of their hired help is difficult to estimate. That so many verses in the *Gazette* were authored by employees and that, periods of dearth aside, the punchers contributed to the paper on a variety of subjects on many occasions, does suggest a degree at least of emotional attachment. It appears to indicate, too, that the *Rocking P Gazette* played a not insignificant role in cultivating that feeling.

Conclusions

In pointing out that the family ranch/farm was better suited to conducting agriculture on the plains than the so-called "great ranches" of an earlier epoch, we have simply looked at the evidence. Through the array of business and personal papers of the Macleay family, including and, above all, the *Rocking P Gazette*, which its descendants have been so careful to retain and preserve, it has been possible here to make that point by examining the conventions and strategies Roderick, Laura, and their daughters embraced. More than anything else, what we have seen is that the Macleays' ability to respond to specific challenges at particular times was one of their most significant overall attributes. Whether it was farming, moving for a time into feeding grain, taking up pork production, or keeping the bank (should one argue) justly shouldering its share of risk, they had the requisite flexibility and hands-on control. Rod Macleay was on his land enough, season in and season out, to see that things were done properly, and Laura, Dorothy, and Maxine were there to help where they could in a wide range of tasks, from gathering eggs to nurturing the garden, shooting game, hunting down stray cattle, or taking care of domestic affairs. Most noteworthy of all was the trust and teamwork Rod and Laura achieved, which ultimately brought order to their financial affairs.

It has been in vogue since World War II to lament the slow disappearance of the family farm/ranch due to poor economies of scale, and its replacement by large corporations that supposedly can take hold of such efficiencies and drive the smaller people out. This has been dramatically overstated. On the one hand, there is no question that vertically integrated giants like Cargill and Tyson Foods have helped to keep many producers financially dependent.[1] On the other hand, all the original ranching corporations failed, and the approach they took has proved insufficient for farming as well as ranching in both the Canadian and American Wests. "The presence of large-scale ownership and giant operations have all been part of American farming throughout the long sweep of time," notes Paul Voisey. "The great slave plantations provide the most conspicuous examples, but the nineteenth-century Midwest also boasted bonanza farms. They arose in the Red River Valley, and even on the Canadian prairies, where mammoth enterprises like the great Bell farm and the Lister-Kaye farms appeared in the late nineteenth century. Because these dinosaurs soon collapsed, historians regarded them as transitory freaks of early western agriculture, but the vision or logic of economies of scale that inspired them survived into the twentieth century."[2] Evidence also suggests that big companies will not play a significant part in the rural western economy in the future. Thus, for instance, One Earth Farms, which formerly rented 200,000 acres, largely from First Nations, has now quit farming altogether in the three prairie provinces. Its CEO admits that the reason is that it did not manage to show "any ability to generate anything remotely resembling profitable numbers."[3]

When it is possible to find accurate records, the difference in efficiency between the great ranches and the family-operated spreads makes it obvious why the big operations have never dominated. The Macleay papers offer a rare chance to juxtapose one family's overall productivity on the second cattle frontier against that of some of the ranch companies on the first. Until closing down operations in 1907, one of those companies, the Walrond ranch, continued to subject a large portion of its cattle to the elements on the open range summer and winter. Unable to watch and guard its cattle closely or to practise seasonal birthing, and being totally dependent on the quasi-devoted hands of hired labour, it experienced a very high death rate. In the earliest period the ranch kept some 3,500 cows for breeding purposes. The highest number of calves it ever recorded was just over 2,400—a birth rate of about 65 percent.[4] The average, however, was about 1,500 calves, or circa 42 percent.[5] The ranch would attempt to keep the steer calves alive for three to five years in order to get them mature, fat, and ready for the beef market. In the 1890s the manager, Duncan McEachran, figured in a death loss of 5 percent per year on those cattle from the time they were weaned from their mother until they were ready for slaughter. However, when proper counts were taken this assumption was always far too low. A more realistic estimate puts it in the range of 20 percent.[6]

The losses on Macleay's ranches were miniscule in comparison. Fences allowed Roderick to keep better track of all the stock year-round and particularly during spring calving season when midwifery was at times a necessity. Having enough roughage and, at times, grain on hand to supply the cattle through the longest winters was always a priority, and fences also made it relatively easy to gather the stock in the fall and then to keep it close to feed and some protection from the elements and from predators when the cold season set in. Moreover, and every bit as important, Rod himself was around most of the time putting his own shoulder to the wheel and seeing that his men were doing the same. He not only regularly achieved a 75 to 77 percent calf crop but also actually did eventually manage to market circa 95 percent of his steer calves thereafter.[7]

The difference between the levels of efficiency achieved by the family outfits and the companies was just as evident in the horse business as it was in the cattle business. Here too it is explained by the ability of the one and failure of the other to give the business the right amount of close attention. As we have seen, Rod Macleay produced much higher-quality beef cattle than the big corporation ranches had been able to generate. In terms of the quality of horses, of the companies, the Quorn ranch near Black Diamond did its best to supply superior mounts for the British army, and the Walrond at one point ran over 600 well-bred Clydesdales and Shires on its rangelands that it hoped would impress the most discriminating buyers in Great Britain.[8] Both failed miserably because they could not manage consistently to produce the type of animals under open rangeland conditions that the very selective Old World buyers wanted. When speaking of the Quorn's breeding program, local rancher Frederick Ings summed up the problem rather succinctly: the "imported mares were not used to rustling on the range, they were not given the care they needed, and though they produced some pretty fair nags, they were not good enough to make … a success."[9]

One partial success that very clearly illustrates the point is George Lane's purebred Percheron operation on the Bar U ranch. Lane started breeding Percherons in 1908, and he produced some fine animals that won numerous awards in horse shows across North America and Europe. Two things need to be understood about Lane's program, however. Firstly, he attained these high standards with only a select few of his horses. All the animals he expected eventually to offer for sale or show were treated with the greatest possible care and attention. To quote the Bar U's modern chronicler, Simon Evans: "In the spring each youngster was carefully inspected, those showing potential being retained as stallion prospects, while the culls were altered and developed as geldings. … Horse colts were grain fed even while at pasture during the summer. They ran in large pastures surrounded by fences of woven wire. Feed bunks were installed in each pasture, in which colts received their daily ration of grain."[10] A series of barns and birthing stalls built on the Bar U home place ensured that the marketable animals could be nurtured and fed indoors and kept in top condition at all times.

In other words, Lane protected the most saleable of his Percherons from the harshest of nature's elements summer and winter. That could not be said about the mares that made up his brood herd. Those animals, like the cattle, he left to fend for themselves most of the time. As Evans reports, "Weaning fillies were well cared for the first year and then turned out on native pasture, receiving no grain from then on. Brood mares were never pampered. They ranged the hills west of the ranch in the summer and were moved to the Bar U flats for the winter [where they] grazed the prairie wool never receiving hay or grain." Lane paid dearly for this part of his program in lost stock. In any reasonably sophisticated breeding program one would expect annual reproduction rates of no less than 75 and as high as 90 percent. In the three years for which Evans was able to find breed books, the Bar U produced respectively eleven foals out of fifty mares in 1912; fifteen foals out of forty-six mares in 1913, and thirteen out of forty-two in 1914. By 1913, nine of Lane's original mares had died, 32 percent had not foaled even once, 52 percent had had only one foal, and 16 percent had had two. In 1915, eighty-four foals were born, seventeen died at birth or soon afterward, one drowned in a slough, and one just disappeared.[11] These are truly dreadful statistics, reflecting, one supposes, poor nutrition during the gestation period—particularly in the wintertime—as well as neglect. They also beyond doubt represent very great financial losses for Lane's horse business as a whole.

The one ranch other than Riddle and Macleay brothers that seems to have made the business work was A. E. Cross's well-known A7 ranch. Cross claimed that it was that side of his operation that was successful enough in the early days to more than make up for calamity on the beef side. After the disastrous winter of 1886/87, it apparently "paid the total capital invested in three years besides 50 head to the good." Two facts need to be considered about Cross's approach, however, that help us understand why Riddle and Macleay brothers and then Macleay on his own found this business worthwhile. First, like them, Cross kept his sights on the local market, which he understood, and he saw that his animals were trained specifically for working cattle and/or hauling and plowing. Second, and probably more importantly, Cross, also like them, dealt in the rougher but sturdy horses that were required mainly by neighbouring ranchers and farmers to work their herds or fields. He "did very well" with these, by "always watching the *local* demand" and having his "horses ready for any purchaser that might come along, and never lost an opportunity of making a sale if any fair price was offered."[12]

Unlike the Walrond, the Oxley, and the Bar U, all of which succumbed to insolvency, Rod Macleay managed to stay in the horse business for many years—indeed, most of his working life.[13] He would not have done so unless it was contributing financially to his business. Today his descendants on the Blades side of the family raise saddle horses, and though their business has changed dramatically, it is still a going concern. This has largely been the result of a change in demand. Draft horses pretty much dropped out of use on grain and mixed farms after World War II, and many ranchers supplement the cattle ponies with all-terrain vehicles. Now another type of horse has slowly gained the acceptance of people who do not want them for any sort of work. "The horse population is astounding," Macleay's grandson wrote just after the turn of the twenty-first century:

> One would think they would have dwindled into oblivion with the advent of mechanization but their population is

now greater than it was when they were an actual necessity. They are still an important part of the cattle business but only a small percentage of the provincial total are used for actual cattle work. It appears today as if the job of the greater majority is to stand in the fence corner to be admired by their owners. The government observed this non-activity and they have been re-classified as a recreational animal instead of a beast of burden. Now even ranches with no ATVs have to admit they use recreational vehicles. They have become indicative of an affluent society.[14]

While it is clear that the family approach to operating ranches or farms on the plains is still the only one with any staying power, it is also undeniable that the size of the average agricultural holding has risen quite a lot in the modern period. This is evident from the various figures provided by the Canadian census reports. From 1961 to 2006, the number of farms in Alberta dropped from over 70,000 to under 58,000, and the average size rose from under 700 to over 1,000 acres.[15]

However, these figures do not suggest the failure of the family farm or its replacement by corporations. They simply indicate that mechanization has enabled families to continue to work more land. Little by little over the decades, starting in the 1920s and even earlier, they slowly utilized the horse less and less and turned to the tractor and a host of other machines to work their fields, harvest grain, and put up roughage. Using the great four-wheel-drive tractors and self-propelled combines, balers, and windrowers, as well as automatic mixing and feeding equipment, it is as easy for a modern operator to sow his crops and look after his livestock on a 1,000-acre unit today as it was for him to do so on a fifth of that in 1931. Presumably attempting to respond to concerns repeatedly articulated about companies taking over the land, the census reporters in 1971 identified all the "incorporated non-family" operations they could find in which controlling interest was "held by shareholders other than the operator and family." They found that a mere 78, or .00124 percent, of the 62,702 farms in Alberta could be described that way.[16]

Much of the story we have told of the historical development of the family ranch as a holistic institution we have directed toward explaining its sustainability in the western foothills of Alberta in the late nineteenth and twentieth centuries. That being the case, it would seem appropriate to conclude the story by owning that the family ranch, ranch/farm, mixed farm, and grain farm have faced and will continue to face significant economic challenges. In the final analysis, to argue that the family unit has traditionally been the only durable means of agricultural production in the Alberta foothills is not to suggest it has been very profitable. One might indeed argue that it has endured largely because it is able to keep going in an industry that tends over much of the time to be uneconomic. The Macleays' case does not challenge that picture. Given the debts that Rod assumed over the years, it took some desperate financial finagling, and the sudden coming together of healthy beef prices and favourable land lease politics at the precise moment when the Gordon, Ironside and Fares firm happened to need cash, to keep them going. Moreover, they survived in part because at that moment in their history they were able to find a wealthy benefactor to finance them so they could take extraordinary steps to protect themselves. In the real world, taking extraordinary steps to counter indebtedness or even just tough times in the agricultural industry is quite, uh … ordinary, in rural Alberta. The members of many country families today take work off the land when in need of cash—she perhaps teaching at a district school or nursing in a local hospital, he doing custom work with his combine or cattle truck or offering guide services to urbanites who wish to experience the outdoors world he knows so well. Such people are acting the way Rod Macleay and partners did when they first settled on the land in the foothills. They are searching for income when their agricultural business is not making enough to pay all the bills. Some rancher/farmers simply work very long days on their own place without what in town would be overtime pay or even just proper hourly recompense to keep the wheels turning. The following is a well-known joke on the northern Great Plains. People on the land invariably laugh at it because there is more than a grain of truth to it.

> Old ranch owner John farmed a ranch in Alberta. The Alberta government claimed he was not paying proper wages to his workers and sent an agent out to interview him.

"I need a list of your employees and how much you pay them," demanded the agent.

"Well," replied old John, "There's my ranch hand who's been with me for 3 years. I pay him $600 a week plus free room and board. The cook has been here for 18 months, and I pay her $500 a week plus free room and board. Then there's the half-wit who works about 18 hours every day and does about 90% of all the work around here. He makes about $10 per week, pays his own room and board and I buy him a bottle of bourbon every Saturday night."

"That's the guy I want to talk to, the half-wit," says the agent.

"That would be me," replied old rancher John.[17]

All this aside, many ranchers and farmers have been able one way or another to achieve a reasonably satisfactory life in the countryside mainly by relying on the family approach, part of which has been to keep all its members pitching in as Laura, Dorothy, and Maxine did whenever and wherever they could. In the case of the latter two, the opportunity to contribute out of doors and thus to blur gender roles, seems actually to have diminished for a while after they started families and then increased again in later years. When they first married and produced their own children (5 Blades and 2 Chattaways), most of their time had to be spent in the domestic sphere. As their children grew older and required less attention, however, they were able once again to devote a greater portion of their energies to working the cattle. In saddle or branding corral, they relied on the skills they had learned as youngsters. Clay Chattaway points out that the model Dorothy and Maxine then reinforced continues to play a crucial part in propagating the ranching industry today. In the late 1940s and early 1950s, he tells us,

> the majority of women were busy with domestic duties. These were numerous and weighty chores that had to be done with scant help from men or clothes-washers, dryers, dishwashers and microwaves. They seldom wore a pair of pants,

drove tractors, rode horses or spent the whole day outside. I would safely venture a guess that they are more now than then, a key component and an integral part of the so-called "agricultural work force." They can accomplish this dressed in women's jeans which were invented after 1950. Men still have clean clothes in the morning, don't dry dishes and have a hot lunch. Many an operation would grind to a halt today without the direct input of women in the corrals, fields and offices. The old adage was women should be bare foot, pregnant and in the kitchen. Today they are well shod, in the branding corrals, at the processing chutes and it is a real nuisance when they are burdened with a pregnancy. ... My Grandfather has 25 of his direct descendants living on his old range. Part of the reason is that the infrastructure is here, so that we can live as comfortable as people in town and attract spouses to come procreate. The other part is a natural social progression. My grandfather had dozens of men working for him. His descendants now replace them all.[18]

One way or another, the system of production that carved out a lasting place for the cattle industry in the foothills of Alberta during the second cattle frontier appears to have gone back to its roots. Now, though, it is more developed infrastructure—including the full array of household conveniences and advanced systems of transportation and communication—rather than a dearth of all such things, that distinguishes it.

We close with a reminder that the *Rocking P Gazette* can now be found online at http://contentdm.ucalgary.ca/digital/collection/rpg.[19]

NOTES

CHAPTER 1

1. Elofson, *Frontier Cattle Ranching in the Land and Times of Charlie Russell* (Montreal: McGill-Queen's University Press, 2004), 13–24.

2. Elofson, *Frontier Cattle Ranching*, 13–24.

3. All types of cattle in all of Alberta and Assiniboia: S. M. Evans, "Stocking the Canadian Ranges," *Alberta History* 26, no. 3 (Summer 1978): 1; Canada, *Fourth Census,* 1901, vol. 2, 51–52.

4. For cowboys and "remittance men," see Elofson, *Frontier Cattle Ranching*, 12–13. The quotation is from a rancher describing the similar situation at the same time across the line in Montana: Granville Stuart, *Forty Years on the Frontier, as seen in the journals and reminiscences of Granville Stuart*, vol. 2, ed. P. C. Philips (Cleveland: A. H. Clark), 35.

5. Canada, *Farming and Ranching in Western Canada* [Montreal, 1890]. The *Daily Courier* in Liverpool claimed that the live cattle trade between Britain and America was so valuable that "anything calculated to curtail its limits could not be regarded as other than a national calamity ("Treatment of Cattle," 15 July 1880); and the Scottish agricultural writer, James Macdonald, told his readers to expect profits in the Trans-Mississippi West to run around 25 percent annually: *Food from the Far West* (New York: Orange Judd, 1878). For the larger story of the promotion of the West on both sides of the forty-ninth parallel, see Elofson, *Frontier Cattle Ranching*, 25–41.

6. S. S. Hall, *Stampede Steve; or, the Doom of the Double Face* (New York: Beadle and Adams, 1884); Prentiss Ingraham, *Buffalo Bill, from Boyhood to Manhood: Deeds of Daring, Scenes of Thrilling Peril, and Romantic Incidents in the Early Life of W.F. Cody, the Monarch of the Borderland* (New York: Beadle and Adams, [1882]); J. S. C. Abbott, *Christopher Carson familiarly known as Kit Carson* (New York: Dodd, Mead and Co., 1874).

7. For the dime and romantic novels, see Elofson, *Frontier Cattle Ranching*, 25–41.

8. London: Shumen Sibthorp, [1902].

9. *The Virginian: A Horseman of the Plains* (New York: Macmillan, 1902). A recent volume sees Everett Johnson, who moved to Alberta in 1888, as the model for Wister's Virginian: John Jennings, *The Cowboy Legend: Owen Wister's Virginian and the Canadian-American Frontier* (Calgary: University of Calgary Press, 2015).

10. H. L. Williams, *The Chief of the Cowboys; or the Beauty of the Neutral Ground* (New York: R. Midewitt, [1870]).

11. *Sky Pilot: A Tale of the Foothills* (Chicago, New York, Toronto: R.H. Revell, 1899).

12. Elofson, *Frontier Cattle Ranching*, 15–24.

13   This migration is vividly depicted in James Belich, *Replenishing the Earth: The Settler Revolution and the Rise of the Anglo World, 1783–1939* (Oxford: Oxford University Press, 2009), 406–17.

14   Lillian Knupp, *Leaves from the Medicine Tree: A history of the area influenced by the tree, and biographies of pioneers and oldtimers who came under its spell prior to 1900* (Lethbridge: High River Pioneers' and Old Timers Association, 1960), 496. James Belich employs the term "settler revolution" throughout *Replenishing the Earth* to describe the spread of English-speaking people and culture to the four corners of the world.

15   We are obviously rather creatively filling gaps in our knowledge of *exactly* how Macleay and the others decided to head west, but we feel quite certain that our narrative cannot be very far off the mark.

16   John and Helen Riddle's son.

17   John's brother.

18   A we will see, John Riddle would buy land and custom feed cattle on their holdings to help them get started.

CHAPTER 2

1   Macleay family papers, Roderick Macleay's diary.

2   NW 32-16-1-W5, #89,479, issued 3 May 1901. Bureaucracy moved slowly in those days too, because it was not until 25 November 1904 that the inspector was there to oversee it. Douglas Riddle got his patent, #112,441, to NE 32-16-1-W5 on 16 July. At the same time Douglas acquired Morrill's patent as per arrangements on SW 32.

3   The Calgary to Fort Macleod leg was also surveyed in 1890, and construction reached Mekastoe/Haneyville (three miles north of Fort Macleod) in 1892. In 1898, a short link with the Crowsnest Pass line was completed. With almost 300 miles of construction complete, the C & E Railway received a total land grant of 1.8 million acres.

4   Macleay family papers, Roderick Macleay's diary.

5   Alex Macleay leased the grazing rights to section 29 to the southeast, and to section 36-16-2-W5 directly to the west of the company's home section, in the name of the partnership. Douglas's father, John Riddle, bought the north half of 34-16-1-W5 in 1904, which he then leased to the company; and when the patent was issued on his homestead on NE 32-16-1-W5 in 1905, he gave it over to Douglas Riddle. On 11 February 1904, Douglas bought section 30 from the Department of the Interior, and a few days later Rod bought another section and a quarter, from the C & E Railway—all of section 25 and the southeast quarter of section 35-16-1-W5. Rod and Douglas rented these parcels to the company (Macleay family papers, Roderick Macleay's diary).

6   Spaying is removing the ovaries. Ranchers in this earlier period, when veterinarians were not readily available, learned to perform the operation themselves.

7   Rod bought out a homestead in 1912 where he set up a winter feeding station they called the calf camp. Weaned calves were sent to the calf camp to be fed over the winter in an area that had protection from the cold winds of winter. It had corrals with good water

and huge stacks of hay. Weak older cows and sometimes breeding bulls were wintered there as well. A married man lived at the calf camp continually so that attention to the cattle could be ongoing (see Charles Walters, below pp. 52, 184).

8   For the spring roundup, see *Rocking P Gazette*, May 1924, 7. It was then that the calves were branded and the males neutered. The females were normally operated on when they were close to a year old.

9   For a fall roundup, see "The Rocking P Round-up," *Rocking P Gazette,* September 1923, 28.

10 Hauling in salt became routine on the Rocking P: see below, pp. 99, 104, 105.

11 These two men are perhaps best known as half of the "Big Four" who underwrote the first Calgary Stampede in 1912. The other two were Patrick Burns and Archibald J. McLean: Warren Elofson, *Cowboys, Gentlemen and Cattle Thieves; Ranching on the Western Frontier* (Montreal: McGill-Queen's University Press, 2000, 157).

12 One child had died soon after birth in 1905.

13 Macleay family papers. For the story of Ware's life in Canada, see Grant MacEwan, *John Ware's Cow Country*, 3rd ed. (Vancouver: Greystone Books, 1995).

14 Warren Elofson, *Somebody Else's Money: The Walrond Ranch Story, 1883–1907* (Calgary: University of Calgary Press, 2009), 175, 180.

15 Another name for young steers or heifers.

16 Elofson, *Somebody Else's Money,* 193–94.

17 Glenbow Archives, New Walrond Ranche papers, M8688-5, David Warnock to Dumcan McEachran, 14 November 1900.

18 "Wholesale market prices for selected agricultural products, 1867 to 1974,"http://www.statcan.gc.ca/pub/11-516-x/sectionm/M228_238-eng.csv.

19 Steers that size will put on at least a pound and a half a day when the weather is good and the grass is lush.

CHAPTER 3

1   Elofson, *Frontier Cattle Ranching in the Land and Times of Charlie Russell* (Montreal: McGill-Queen's University Press, 2004), 2–24.

2   Elofson, *Somebody Else's Money: The Walrond Ranch Story, 1883–1907* (Calgary: University of Calgary Press, 2009), 141–53.

3   Elofson, *Cowboys, Gentlemen and Cattle Thieve; Ranching on the Western Frontiers* (Montreal: McGill-Queen's University Press, 2000), 27–29, 48–49, 147. Some western breeders, who closely monitored the local market, also mated their cayuse mares with Percheron or even Clyde or Shire stallions to give them still greater size and strength.

4   "Wholesale market prices for selected agricultural products, 1867 to 1974," http://www.statcan.gc.ca/pub/11-516-x/sectionm/M228_238-eng.csv.

5   Max Foran, *Trails and Trials: Markets and Land Use in the Alberta Beef Cattle Industry, 1881–1918* (Calgary: University of Calgary Press, 2003), 32.

6   Foran, *Trails and Trials*, 32.

7   Emerson made mention of them in a letter to Rod dated 21 December 1902: "Am glad to hear you had a good trip and got back safe. I wanted to see the horses you bought. I am off to Mexico, good bye to all until I get back." Demand for horses also apparently came from the British army, particularly during the Boer War (1899-1902): "Horses for South Africa," *Macleod Gazette*, 21 February 1902.

8   Macleay family papers.

9   Armour was an American meat-packing company started in 1867 in Chicago, by the Armour brothers under the lead of Philip Danforth Armour. Later, its plant in Omaha, Nebraska, gave that city the largest meat-packing industry in the United States.

10 Joseph Harrison, owner of the 7U ranch near Pekisko Creek; see Glenbow Archives, M6552: "Margeret Barry-McGeche,'7U' Brown, Alberta Cattle Company, 1881-1883, prepared for Alberta Culture, c. 1985."

11 Macleay family papers, Roderick Macleay's diary.

12 A "settlers rail car with effects including 7 horses, owner risk, weighing 24000 lbs @60.00, Barge Rental $40.00" gives an idea of what transportation fees from High River to Revelstoke were (Macleay family papers). Their bill for hay and shipping to St. Leon's in January 1908 was $175.18. R. L. McMillan homesteaded in Alberta on section 11, Township 14, Range 3, Meridian 5 (167671-167680-Alberta Homesteads 1830-1970, file 70136).

13 Simon M. Evans, *The Bar U and Canadian Ranching History* (Calgary: University of Calgary Press, 2004), 134-37.

14 E. C. Abbott and H. Huntington Smith, *We Pointed them North: Recollections of a Cowpuncher*, 2nd ed. (Norman: University of Oklahoma Press, 1955), 176.

15 L. V. Kelly, *The Range Men*, 75th anniversary ed. (Calgary: Glenbow-Alberta Institute, , 1988), 191.

16 Elofson, *Cowboys, Gentlemen and Cattle Thieves*, 85.

17 Provincial Archives of Alberta, Edmonton, 72,27/SE: Violet LaGrandeur, "Memoirs of a Cowboy's Wife," 5.

18 Kelly, *The Range Men*, 191; Elofson, *Somebody Else's Money*, 214-220.

19 Canada, *Sessional Paper* 42, no. 14 (1907-1908), n 28, 56: Annual Report for D division, 1 November 1907.

20 Hazel Bessie Roen, *The Grass Roots of Dorothy, 1895-1970*, 2nd ed. (Calgary: Northwest Printing and Lithographing, 1971), 105.

21 See letters between Cross and Douglas, through January, February, and March 1907, Glenbow Archives, Cross papers, M1543, f. 470-77.

22 Cross papers, f. 470: Douglass to Cross, 20 January 1907. See also Roen, *The Grass Roots of Dorothy*, 105.

23 Cross papers, f. 470: Douglas to Cross, 27 January 1907.

24 Cross papers, f. 471: Douglass to Cross, 16 March 1907.

25 Macleay family papers. For more on A. E. Cross's experience on the Red Deer, see Elofson, *Cowboys, Gentlemen and Cattle Thieves*, 88–90.

26 "Lost," *Rocking P Gazette*, November 1923, 46.

27 Macleay family papers.

28 Dorothy Margaret, born 26 January 1909.

29 Please see the discussion of race in chapters 11 and 13.

30 Macleay family papers.

CHAPTER 4

1 Prices actually climbed generally. Nationally, steers averaged $.0485/pound in 1907 and $.0829 in 1914. "Wholesale market prices for selected agricultural products, 1867 to 1974," http://www.statcan.gc.ca/pub/11-516-x/sectionm/M228_238-eng.csv.

2 Lillian Knupp, *Leaves from the Medicine Tree: A history of the area influenced by the tree, and biographies of pioneers and oldtimers who came under its spell prior to 1900* (Lethbridge: High River Pioneers and Old Timers Association, 1960), 496.

3 Knupp, *Leaves from the Medicine Tree*, 23.

4 The trail east was originally thought to be a one-way trip to grass gains and then on to market, but after that winter (1914) it became a two-way trip, with many head coming home to the hills for winter. Besides the hay that was put up at the home ranch, more was purchased. It was not transported home, but rather the cattle were farmed out and fed where the hay was located (Macleay family papers, Roderick Macleay's diary).

5 Clay Chattaway comments: "At the Home place, the same was done for years." Macleay's sons-in-law, George Chattaway and Ernie Blades, continued the practice after they took over the Alberta ranchlands in the 1950s.

6 "Local News," January 1925, 6; March 1924, 2.

7 "Local News," February 1924, 9; "Cartoons," March 1925, 9.

8 See below, p. 71.

9 "Local News," November 1923, 4.

10 "Local News," November 1923, 4. The *Gazette* of September 1924 notes that a hired hand named "Ezra Left-hand" and a "company" of assorted other men "finished stooking the wheat at the home-place" ("Local News," 8).

11 "Local News," October 1924, 2.

12 "Local News," 6.

13 See, for instance, "How Do Fertilizers Affect the Environment?" https://helpsavenature.com/how-do-fertilizers-affect-environment.

14 "Roderick MacLeay – Pioneer Rancher," *Lethbridge Herald*, 12 December 1931. "Feedlots" were corrals designed to feed cattle usually to get them ready for market.

15 "Roderick MacLeay – Pioneer Rancher." See also Macleay family papers, "Beef Exports to the United Kingdom, 1930–1931–1932–1933."

16 A note in "Sales to Great Britain, 1930 – 1933," Macleay family papers, reads: "At last it looks as if all Rod's work on the Council of Beef Producers, the W.S.G.A. and his personal push on all Dept of Agriculture personnel has begun to pay off."

17 A. C. Rutherford, "The Cattle Trade in Western Canada," quoted in Kelly, *The Range Men*, 75th anniversary ed. (Calgary: Glenbow-Alberta Institute, 1988), 200.

18 "Yesterday's Market," *Edinburgh Courant*, 18 September 1885.

19 "Yesterday's Markets," *Edinburgh Courant*, 12 September 1885.

20 "Calgary Bull Sale Nets New Records," *Farm and Ranch Review*, 1 April 1947.

21 Clay Chattaway.

22 "Local News," April 1925, 1. As late as 1947 Rod bought the Champion bull at the Calgary bull sale ("Calgary Bull Sale Nets New Records," *Farm and Ranch Review*, 1 April 1947).

23 Clay Chattaway.

24 "Sales to Great Britain, 1930–1933."

25 Warren Elofson, "Grasslands Management in Southern Alberta: The Frontier Legacy," *Agricultural History* 86, no. 4 (2012): 143–68. This also evinces Macleay's recognition of the necessity of looking after his pasturelands. The only way to produce fully finished steers on grass was and is to keep pastures lush, voluminous, and healthy, mainly by being careful never to overgraze them and by pulling all the cattle off them during extended periods each year.

26 "Letter from Ottawa, Dec 5, 1930 from Office of Deputy Minister of Agriculture, Robert Weir, to RRM."

27 "Sales to Great Britain, 1930–1933."

28 "High River Rancher Urges National Marketing Scheme Before Stevens Commission," *Lethbridge Herald*, 22 March 1934.

29 The Hawley-Smoot tariffs were rescinded in 1936.

30 Warren Elofson, *Somebody Else's Money: The Walrond Ranch Story, 1883-1907* (Calgary: University of Calgary Press, 2009), 21–62.

31 Simon M. Evans, *The Bar U and Canadian Ranching History* (Calgary: University of Calgary Press, 2004), 193–218.

32 As to the purchase of the TL, see chapter 6.

33 See chapter 8.

34 "Local News," April 1925, 1.

35 "Local News," April 1925, 3

36 "Local News," April 1925, 2. Charles Walters lived at the calf camp with his wife and at least one daughter and with Jesse Walters, who must have been a son. The calf camp

was where the calves were kept after being weaned in the fall. The calves were kept there through the entire winter so they could be fed and looked after.

37   "Local News," December 1924, 4.

38   "Local News," February 1924, 2. Bill Livingstone had a ranch at Willow Creek.

39   "Local News, February 1925, 3. "Drumhellar's" refers to the homestead of Dan Drumhellar, who was a cowhand on Macleay ranches.

40   "Local News," October 1924, 7. Tex Smith, cowpuncher. Peddie unknown.

41   "Local News," October 1924, 7.

42   "Local News," September 1924, 3.

CHAPTER 5

1   Elliott West, "Families in the West," *Organization of American Historians Magazine of History* 9, no. 1 (Fall 1994): 18–21.

2   Sarah Carter observes that while some works, such as those by Sheila Jameson and Lewis Thomas, examine the lives of women on western ranches, "a cherished myth of an entirely masculine ranching culture and cattle industry has proven difficult to dislodge." "'He Country in Pants' No Longer—Diversifying Ranching History," in *Cowboys, Ranchers and the Cattle Business: Cross Border Perspectives on Ranching History*, ed. Simon Evans, Sarah Carter, and Bill Yeo (Calgary: University of Calgary Press, 1997), 155; Shelagh S. Jameson, "Women in the Southern Alberta Ranch Community, 1881-1914," in *The Canadian West: Social Change and Economic Development*, ed. Henry C. Klassen (Calgary: University of Calgary Comprint Publishing, 1977), 63–78; Lewis G. Thomas, *Rancher's Legacy: Alberta Essays*, ed. Patrick E. Dunae (Edmonton: University of Alberta Press, 1986). The best source for women's contributions to the family ranch in the western foothills is Rachel Herbert, *Ranching Women in Southern Alberta* (Calgary: University of Calgary Press, 2017). For a paralleling of nature and gender issues, see Carolyn Merchant, *Reinventing Eden: The Fate of Nature in Western Culture* (New York: Routledge, 2003), 93–143. An extensive historiography of farm women exists; see also Linda Rasmussen et al., eds., *A Harvest Yet to Reap: A History of Prairie Women* (Toronto: The Women's Press, 1976); Veronica Strong-Boag, "Pulling in Double Harness or Hauling a Double Load: Women, Work and Feminism on the Canadian Prairie," in *The Prairie West: Historical Readings*, ed. R. Douglas Francis and Howard Palmer, 2nd ed. (Edmonton: University of Alberta Press, 1992), 401–23.

3   Macleay family papers, Roderick Macleay's diary.

4   *A Female Economy: Women's Work in a Prairie Province* (Montreal: McGill-Queen's University Press, 1998), 95.

5   Gertrude Lillian Sturtevant was born on 19 May 1882, and her sister Laura Marguerite on 12 December 1884, in Burlington, Chittenden County, Vermont, to Charles and Adelaide [Frenier] Sturtevant. Two boys, Curtis and Wallace, were added to the family. Charles Sturtevant, Laura's father, died in 1891. At about the same time, a fire destroyed their home, leaving Adelaide and four young children homeless. Charles's sister

Hortense and her husband Robert Wright, a banker, took the girls Laura and Gertrude to raise. It was hard for Adelaide to let them go but fortunate for them that she did, as they both benefited from a superior education and a fine home, something that the boys were not able to obtain. Dorothy was also educated at Newport Academy, Newport, Vermont (Clay Chattaway).

6   "Local News," *Rocking P Gazette*, October 1923, 1: "On the 2nd day of this month Clem Hensen, trailed to the trading post of Cayley, with the double wagon, and brought back the winter's grub-stake for the Rocking P outfit."

7   Macleay family papers, Clay Chattaway, "Ranching Changes 1945 to 2005."

8   "Local News," *Rocking P Gazette*, February 1924, 6: "Wick [LeMaster] and Tommy [MacKinnon] put up the first load of ice on Feb. 28th."

9   A stone boat was a flat wooden platform on skids used for moving heavy objects. It was horse-drawn.

10   P. 29

11   "Local News," 8.

12   Macleay family papers, "High River, Alberta, 20th February, 1919, Names of men employed by Roderick R. Macleay during 1918." For vehicles as well as roads, see below, present chapter and chapter 12.

13   "Lost and Found," September 1924, 67.

14   "Local News," 6.

15   "Local News," September 1923, 7.

16   "Local News," 5.

17   "Local News," January 1925, 6.

18   "Local News," September 1924, 8.

19   "Local News," February 1924, 4.

20   "Local News," January 1925, 1.

21   "Local News," January 1924, 5.

22   "Local News," January 1924, 7.

23   A cigarette.

24   That is, the Rocking P Ranch.

25   October 1924, 25.

26   Glenbow Archives, Cochrane papers, M65552-2: Evelyn Cochrane to Arthur Cochrane, 29 November 1900. I am indebted to Rachel Herbert for this reference and for those in notes 27–30 below: see *Ranching Women in Southern Alberta*, 172–73.

27   Pedersen family, in Pincher Creek Historical Society, *Prairie Grass to Mountain Pass* (Pincher Creek: Pincher Creek Historical Society, 1974), 797.

28   Bateman family, in Foothills Historical Society, *Chaps and Chinooks: A History West of Calgary* (Calgary: Foothills Historical Society, 1976), 257.

29 Halton family, in Pincher Creek Historical Society. *Prairie Grass to Mountain Pass*, 67.

30 *Prairie Grass to Mountain Pass*, 261–62.

31 Warren Elofson, *So Far and Yet So Close: Frontier Cattle Ranching in Western Prairie Canada and the Northern Territory of Australia* (Calgary: University of Calgary Press, 2015), 192.

32 "A Remarkable Man. The Late Mr. W.P. Hayes, Sen.," *The Register*, 19 November 1913. William figured his daughters were mentally strong too. "I can tell you that if they were bosses of a station things would have to be carried out their way."

CHAPTER 6

1 Macleay family papers, "John Ware Ranch," compiled history, handwritten notes [date and author unknown]. Clay Chattaway comments: "Emerson got some of the money but it was mostly put towards paying off the bank for the cattle. Emerson was paid in instalments as was usual. He received the final payment in July 1915."

2 Formerly owned by Paddy Langford.

3 Macleay family papers, Roderick Macleay's diary.

4 As opposed to private land such as that of the CPR or Hudson's Bay Company.

5 Clay Chattaway comments: "These were the days when a cow was worth more than an acre of land and it was somewhat easier to make a land purchase pencil out. But it was not so much a situation of whether or not the land would pay for itself, but as to whether or not one had the money, or had the ability to swing the credit. The real profits at this time were in grain and the demand was strong for farmland, the opposite was true of ranch land."

6 Burns had previously acquired the lease from the well-known Conrad brothers of Montana.

7 10 April 1919.

8 Macleay family papers.

9 Glenbow Archives, Burns papers, M160-231: P. Burns Ranches Ltd. to Riddell, Stead, Graham & Hitchison (now KPMG), 18 March 1924. M160-231: P. Burns Ranches Ltd. to Bank of Montreal, 3 January 1922.

10 "Wholesale market prices for selected agricultural products, 1867 to 1974," http://www.statcan.gc.ca/pub/11-516-x/sectionm/M228_238-eng.csv.

11 Macleay family papers, "Burns and Macleay Law Suit."

12 Judge Walsh stated: "My conclusion upon the whole is that the contract, is that set up by the plaintiff. Defendant's action to sell to him only circle cattle on circle ranch must fail," Macleay Family Papers, "Burns and Macleay Law Suit."

13 R. B. Bennett charged $3,871.76 for the Macleay v. Burns lawsuit over the Circle cattle and agreed to monthly payments of $200, Macleay Family Papers, "Burns and Macleay Law Suit." Burns turned the Circle 3 brands over to Macleay Ranches.

14 It was possible to appeal to the Judicial Committee of the Privy Council until 1949.

15 The judge had ordered Macleay to "pay $0.50 per head per month to Burns" for the care and feed of the cattle "since 9 June," when the original deal was struck.

16 Macleay family papers, quoting Maxine Macleay.

17 Macleay family papers, Dorothy Macleay, compiled history [date unknown].

18 Clay Chattaway.

19 And an audit for the year 1924 showed a loss of $27,000.00 (Macleay family papers, "1924," compiled history [date and author unknown]).

20 Glenbow Archives, Burns papers, M160–202: Riddell, Stead, Graham & Hutchison to P. Burns, 13 April 1925.

21 "1924 per head value of all non-dairy cattle in Canada: 27.11," online: http://www.statcan.gc.ca/pub/11-516-x/pdf/5220015-eng.pdf.

22 Since Rod paid $90 for the Burns cattle we assume that they were all older steers.

23 Burns papers, M160–202: R. Macleay to E. Corlet, 11 April 1925.

24 Macleay family papers: Bank of Montreal to Rod Macleay, 4 August 1923.

25 As opposed to today's standards, which are asset (or land) based.

26 The original brand was 4 walking sticks; one was dropped, making 3, which looked like 9s opened up so as not to blotch. Some called it the "three 7s" (Clay Chattaway).

27 Burns papers, M160–231: Bank of Montreal to P. Burns Ranches Ltd., 22 December 1921; M160–231: Letter to Macleay from Burns, 7 December 1921.

28 "76" was the brand of the Powder River Cattle Co. taken over by Gordon, Ironside and Fares. For the Gordon, Ironside and Fares story, see A. B. McCullough, "Winnipeg Ranchers: Gordon, Ironside and Fares," *Manitoba History* 41 (Spring/Summer 2001): 18–25.

29 A rancher from near Winnipeg named Harvey Iliff Wallace was the president and also a shareholder. The two founders of Gordon, Ironside and Fares Company Ltd. were James T. Gordon and Robert Ironside Senior: Manitoba Historical Society website at http://www.mhs.mb.ca/docs/mb_history/41/winnipegranchers.shtml.

30 Burns papers, M160–202: Mule Creek Cattle Co. 1924–26; M160–202: "Agreement" dated 9 August 1924; Macleay family papers, "Rod Macleay's Purchase," compiled history [date and author unknown].

31 There were 340 cows, 437 steers two years and up, 250 two- and three-year-old heifers, 75 yearling steers and heifers, and 35 saddle horses. See Burns papers, M160–202: agreement dated 9 August 1924; Macleay family papers, "Rod Macleay's Purchase," compiled history [date and author unknown].

32 The terms are evident in the written agreement (Burns papers, M160–202: agreement dated 9 August 1924).

33 When he bought the TL.

34 Max Foran, *Trails and Trials: Markets and Land Use in the Alberta Beef Cattle Industry, 1881–1918* (Calgary: University of Calgary Press, 2003), 130–35. In Alberta, farmers were agitating to have them either thrown open to settlement or turned into community pastures. Community pastures never really gained ground until the Prairie Farm Rehabilitation Act, enacted on 17 April 1935, which created the Community Pasture Program. James Gray, *Men Against the Desert* (Saskatoon: Western Producer Prairie Books, 1967), 169.

35 Macleay family papers, compiled history [date and author unknown].

36 Riley was also founding president of the High River Local of the United Farmers of Alberta.

37 In 1925: Foran, *Trails and Trials*, 134–35.

38 Macleay family papers.

39 Much to the Macleays' benefit, this would prove correct in 1925 when all western Canadian leases that could meet the criterion of agricultural insufficiency were to be given the requested much longer terms (note 37 above).

40 On 5 June 1923, Burns wrote to the provincial treasurer of Manitoba, Hon. F. M. Black, asking if he could put a word in with someone who might be willing to loan Macleay money on some 15,000 acres. "The land is all good land, not ordinary ranch property, most of it being fit for mixed farming. He tells me he would require about $100,000." (Burns papers, M160-233.)

41 Warren Elofson, "Patrick Burns," Dictionary of Canadian Biography, http://www.biographi.ca/en/bio/burns_patrick_16E.html.

42 In the 1930s, Burns was actually appointed to the Canadian Senate by the Liberal government.

43 Burns papers, M160-215: "Schedule C." Some of the horses were branded with the O and some with the 76 brand.

44 "Local News," November 1923.

45 Macleay family papers, compiled history [author and date unknown]. It also states that "370 head of L4L (Streeter) cattle and 456 more home ranch cattle were [also] sent. 500 calves were bought from Mitchell brothers at Medicine Hat and 257 steers from Hardwick."

46 Burns papers, M160-215: P. Burns Ranches Ltd. to Macleay, 28 August 1928. On 28 August 1928, Macleay paid Burns $46,247.91, "representing payment in full of our indebtedness against Mrs. Laura Macleay covering cattle situated on Ranch '76'." This would have been possible because of the RBC loan. See also Letter to Laura from A. Lewis, Manager of Royal Bank of Canada, 22 August 1928 (Macleay family papers).

47 Burns papers, M160-202: "Schedule A."

48 "Livestock statistics, number on farms and farm values at 1 June, Canada, 1906 to 1975," and "Wholesale market prices for selected agricultural products, 1867–1974," http://www.statcan.gc.ca/pub/11-516-x/pdf/5220015-eng.pdf.

49 Burns papers, M160-215.

50   Macleay family papers, compiled history [author and date unknown].

51   Burns papers, M160-215: Western Ranching Limited to the North of Scotland Canadian Mortgage Co. Ltd., 27 November 1929.

52   Calculated by Clay Chattaway.

53   Burns papers, M160-215: P. Burns to C.W. Chesterton, Superintendent, Bank of Montreal, 2 February 1929. Clay Chattaway comments: "The O is the primary brand because it is on the rib, the 3 is secondary and only used because the O is too easy to tamper with. They were always used together but were registered separately."

54   Macleay family papers, Dorothy Macleay compiled history [date unknown]. In fact, they started their own cattle herd as early as January 1924: "Dorothy and her 'pard' started in the cow business this month. Max got a heifer ... Dorothy got a steer ..." (*Rocking P Gazette*, "Local News," 6).

55   "The Bank of Montreal in Calgary were after Dad to give them a [quit] claim deed to all land and cattle; Dad suggested to G.W. Spinnery at the Head Office in Montreal that he [Rod] go down and negotiate a settlement with the men in Head Office as the relations with their men in Calgary 'were strained to say the least.' Spinney replied that 'any negotiations should be conducted through our local representatives and it would not suit us therefore, to negotiate a settlement of your debt thru one of our Head officials.' Dad went to Montreal in February but still made no headway towards a settlement of the ranch debt. Alexander Hannah, a canny old Scot in Bennett's law firm, was the lawyer and he kept up a barrage of proposals and ideas with the Bank's lawyers McLaws and McLaws." (Macleay family papers, Dorothy Macleay, compiled history [date unknown]).

56   "File Defence in Suit for $370,000," *Lethbridge Herald*, 11 September 1941.

57   He had done this once before. In 1925, he negotiated with the Bank of Montreal and it lowered his interest on his bank loan by $10,657.69 and reduced his overall indebtedness by about 109,000.00, (Macleay family papers, Dorothy Macleay, compiled history [date unknown]).

58   See John Feldberg and Warren Elofson, "Financing the Palliser Triangle, 1908-1913," *Great Plains Quarterly* 18 (Summer 1998): 257-68.

59   P. 3.

60   See, as well, David C. Jones, "An Exceedingly Risky and Unremunerative Partnership: Farmers and the Financial Interests Amid the Collapse of Southern Alberta," in *Building Beyond the Homestead*, ed. David C. Jones and Ian MacPherson (Calgary: University of Calgary Press, 1985), 207-27.

61   Kathryn McPherson, "Was the Frontier Good for Women? Historical Approaches to Women and Agricultural Settlement in the Prairie West, 1870-1925," *Atlantis: A Women's Study Journal* 25, no. 1 (2000): 80.

62   "Local News," 5.

63   See also "Local News," May 1924, 5; November 1924, 2; October 1924, 2.

64   Dee Garceau-Hagen, *The Important Things of Life: Women, Work, and Family in Sweetwater County, Wyoming, 1880-1929* (Lincoln: University of Nebraska Press, 1997).

65 "Old Stock and Other Notes," *The Northern Miner*, 2 December 1905.

66 McPherson, "Was the Frontier Good for Women?," 79.

67 In 1947. "Wholesale market prices for selected agricultural products, 1867–1974," http://www.statcan.gc.ca/pub/11-516-x/pdf/5220015-eng.pdf.

68 Ibid.

69 Henry C. Klassen, "A Century of Ranching at the Rocking P and Bar S," in *Cowboys, Ranchers and the Cattle Business: Cross Border Perspectives on Ranching History*, ed. Simon Evans, Sarah Carter, and Bill Yeo (Calgary: University of Calgary Press, 1997), 112–13.

CHAPTER 7

1 January 1924, 17.

2 December 1924, 15.

3 May 1924, 53, 61.

4 Paul Voisey, *Vulcan: The Making of a Prairie Community* (Toronto: University of Toronto Press, 1988), 175–98.

5 April 1924, 8. Clay Chattaway mentions the Macleays' fear of the Spanish Flu.

6 In 1918 and 1919 they had a teacher by the name of Mrs. Estella Gokey for grades three and four. Then, from 1920 to 1921, their teacher was Miss Elizabeth Moore. Dorothy notes she was "a very nice lady, and a good teacher. She was English and liked to ride. Her fiancé had been killed in WWI and she never married" (Macleay family papers). By the fall of 1922 the flu had subsided and Laura moved with Maxine and Dorothy to Calgary for the school year, where they could attend Earl Grey Elementary. In the winter of 1925/26 the sisters went to Anna Head School for Girls, in Berkeley, California, boarding in residence, and the next year the girls were back in Calgary at St. Hilda's. Dorothy then returned to Anna Head in California in 1927–28 while Maxine went to St. Hilda's as they both finished high school. Dorothy and Maxine went on to the University of Alberta late in 1929.

7 "Scene, Rocking P Kitchen," November 1923, 27.

8 Respectively, pp. 37 and 23.

CHAPTER 8

1 "Local News," January 1924, 2.

2 Rod and Laura Macleay travelled in their own car by 1924 ("Local News," October 1924, 6). By this time, the family operations were beginning to haul grain in what would now be considered relatively small trucks (November 1924, 24).

3 "Local News," January 1924, 2.

4 "Visit to the E.P. Ranch," September 1923, 5.

5 February 1925, 12. The workers were Jim Hendrie and Bob Reeves.

6 Credited to "Homer Milton" (but in Maxine's hand), February 1924, 45–46.

7 Warren Elofson, *Somebody Else's Money: The Walrond Ranch Story, 1883-1907* (Calgary: University of Calgary Press, 2009), 113–35; Warren Elofson, *Cowboys, Gentlemen and Cattle Thieves: Ranching on the Western Frontier* (Montreal: McGill-Queen's University Press, 2000) 134–49.

8 May 1924, 68–69.

9 January 1924, 35–39.

10 January 1924, 35–36. See also E.B.W., "The Foothill Country," May 1924, 35.

11 *Wild Life, Land and People: A Century of Change in Prairie Canada* (Montreal: McGill-Queen's University Press, 2016), 252–89.

12 Elofson, *Somebody Else's Money,* 161–62

13 Wetherell, *Wild Life, Land and People,* 356–57.

14 P. 63.

15 Clay Chattaway.

16 Clay Chattaway, "Ranching Changes 1945 to 2005," Macleay Family papers.

17 Michael McGerr, *A Fierce Discontent: The Rise and Fall of the Progressive Movement in America, 1870-1920* (New York: Free Press, 2003), 105.

18 William L. Bowers, *The Country Life Movement in America, 1900-1920* (Port Washington, NY: Kennikat Press, 1974), 3–4.

19 Bowers, *Country Life,* 28.

20 David C. Jones, "'There Is Some Power About the Land': The Western Agrarian Press and Country Life Ideology," in *The Prairie West: Historical Readings,* ed. R. Douglas Francis and Howard Palmer (Edmonton: University of Alberta Press, 1992), 457.

21 Pp. 27–33.

22 Edward C. Abbott and Helen Huntington Smith, *We Pointed Them North: Recollections of a Cowpuncher,* 2nd ed. (Norman: University of Oklahoma Press, , 1955), 223.

23 P. 33.

24 Pp. 23–30. Van Eden (sometimes spelled Eeden), became a cowpuncher on the Bar S.

25 *Census of Prairie Provinces: Population and Agriculture,* 1926. xii, http://publications.gc.ca/collections/collection_2017/statcan/CS98-1926.pdf.

26 In 1926, Calgary had a population of 65,291. Historians have tended to overlook the longevity of ranching culture in the foothills region. For instance, John Herd Thompson, *Forging the Prairie West: The Illustrated History of Canada* (Oxford: Oxford University Press, 1998), delves into economic and social history, stressing the importance of the region and giving only passing reference to the ranching industry.

27 The first "talkie," *The Jazz Singer,* was released in October 1927.

28 The Rocking P brand looks somewhat like an anchor.

29 Ed Orvis also did work such as driving field equipment, including binders.
30 Robert Raynor (ranch carpenter and also a justice of the peace).
31 Tex Smith, cowpuncher.
32 Stewart Riddle.
33 The Bar S.
34 Jim Hendrie (also known as Highland Jim), cowpuncher.
35 Probably Bob Reeves, ranch hand; Tom McKinnon; and Donald Comrie, ranch hand who homesteaded on a ranch south of the Rocking P.
36 Val Blake, cowpuncher.
37 Cowpuncher who died in 1925. The *Gazette* spells his name Kreps, Krepps, and Creps.
38 Ranch hand at the Bar S. In the *Gazette,* they switch between spelling his name Eden and Eeden.
39 "Birth of the Hollywood Cowboy, 1911," blog post, http://www.eyewitnesstohistory.com/hart.htm (2006).
40 "Gilbert M. 'Broncho Billy' Anderson (1881–1971) [was] the genre's *first* western film hero and star, who made about 400 'Broncho Billy' westerns, beginning with *Broncho Billy and the Baby* (1910); his last *silent* western role was in *The Son of a Gun* (1919)": (Tim Dirks, AMC filmsite blog, Western Films, pt. 2 at http://www.filmsite.org/westernfilms2.html).
41 "Hoot Gibson," https://en.wikipedia.org/wiki/Hoot_Gibson.
42 "Tom Mix (1880–1940) … [was] a prototypical western action hero with a wholesome screen persona, fancy cowboy outfits, and his horse Tony the Wonder Horse, a prominent star for Fox films.… He was known as the first western superstar, and first appeared as Bronco Buster in Selig Polyscope's *Ranch Life in the Great Southwest* (1910), and then in many others (for Selig and later for Fox), including *The Man From Texas* (1915), *The Heart of Texas Ryan* (1916), and later in such expensive features as Fox's *Riders of the Purple Sage* (1925) … and *The Great K & A Train Robbery* (1926)," (Tim Dirks, AMC filmsite blog, Western Films, pt. 2: http://www.filmsite.org/westernfilms2.html).
43 "Buck Jones," https://en.wikipedia.org/wiki/Buck_Jones.
44 Dominique Brégent-Heald, *Borderland Films; American Cinema, Mexico, and Canada during the Progressive Era* (Lincoln: University of Nebraska Press, 2015), 204.
45 "Current Empress Attractions," *Macleod Times,* 14 July 1921. For the movie industry in High River, see Paul Voisey, *High River and the Times: An Alberta Community and its Weekly Newspaper, 1905–1966* (Edmonton: University of Alberta Press, 2004), 137–38.
46 "Impersonating Bandit Lands, Hoot in Cell," *Macleod Times,* 31 January 1924; "Hoot Gibson Comes to Empress in New Western," *Macleod Times,* 17 January 1924; "The Empress Theatre Current Attractions: Hoot Gibson Offers a New Characterization," *Macleod Gazette,* 21 September 1922; "Buck Jones Coming in New Fox Picture," *Macleod Times,* 12 January 1922.

47 Probably Ropeswift Ralph—a ranch hand who lived with his wife at the Rocking P.

48 Clem Henson, "The Wild Buckaroo," May 1924, 57. Henson was a Rocking P cowpuncher.

49 Cowpuncher.

50 "Wicked Wick," teamster.

51 "The Rocking P Round-up," September 1923, 28.

52 A prolapse is what veterinarians call the outward collapse of the rectum or vagina. Normally these are pushed back in and stitched up.

53 "Antelope Al" (Maxine), September 1923, 30.

54 I.e. by grabbing it by the horns and throwing it to the ground, often by jumping on it from the back of a fast horse.

55 See L. V. Kelly, *The Range Men*, 75th anniversary ed. (Calgary: Glenbow-Alberta Institute, 1988), 216–220.

CHAPTER 9

1 Edward C. Abbott and Helen Huntington Smith, *We Pointed Them North: Recollections of a Cowpuncher*, 2nd ed. (Norman: University of Oklahoma Press, 1955), 101.

2 *We Pointed Them North*, 87.

3 *We Pointed Them North*, 222–33; E. J. "Bud" Cotton, *Buffalo Bud: Adventures of a Cowboy* (North Vancouver: Hancock House, 1981), 45.

4 *We Pointed Them North*, 223.

5 *We Pointed Them North*, 231–22.

6 Edith Fowke, "American Cowboy and Western Pioneer Songs in Canada," *Western Folklore* 21 (1962): 247–56.

7 "The Cowboy's Life," in *Songs of the American West*, ed. R. E. Lingenfelter, R. A. Dwyer, and D. Cohen (Berkley and Los Angeles: University of California Press, 1968), 347.

8 R. Lithicum, "The Rough Rider" (1895), quoted in *Trailing the Cowboy: His Life and Lore as told by Frontier Journalists*, ed. C. P. Westermeier (Caldwell, ID: The Caxton Printers, 1955), 269–70.

9 Warren Elofson, *So Far and Yet So Close: Frontier Cattle Ranching in Western Prairie Canada and the Northern Territory of Australia* (Calgary: University of Calgary Press, 2015), 191–205.

10 Abbott and Huntington Smith, *We Pointed Them North*, 211.

11 Warren Elofson, *Cowboys, Gentlemen and Cattle Thieves: Ranching on the Western Frontier* (Montreal: McGill-Queen's University Press, 2000), 32. From the 1890s on, far more Americans flowed into Alberta than even Britons.

12 Ranch carpenter and justice of the peace.

13 May 1924, 57.

14 March 1925, 57. Hendrie, also known as "Highland Jim," was a cowpuncher.

15 October 1923, 49–50, credited to Ropeswift Ralph (pseudonym), shows how concerned Macleay cowpunchers were in the 1920s about the danger of a stampede.

16 November 1923, 45. See also Warren Elofson, *Frontier Cattle Ranching in the land and Times of Charlie Russell* (Montreal: McGill-Queen's University Press, 2004), 98–104.

17 November 1923, 45.

18 April 1924, 57–58.

19 For one, Tommy McKinnon (November 1924, 4).

20 April 1925, 73.

21 May 1924, 10.

22 October 1924, 82.

23 Charles Russell, *Trails Plowed Under: Stories of the Old West,* introduction by W. Rogers and B. W. Dippie (Lincoln: University of Nebraska Press, 1996), 51.

24 Elofson, *Frontier Cattle Ranching,* 104–5.

25 See Paul Voisey, *Vulcan: The Making of a Prairie Community* (Toronto: University of Toronto Press, 1982), 158–74.

26 The Leman family homesteaded west of the home place and later started up an operation near Muirhead school (Clay Chattaway).

27 All local ranchers.

28 "Local News," February 1924, 7–8.

29 For a discussion of this subject, see Elofson, *Frontier Cattle Ranching,* 42–62.

30 Below, pp. 163, 164.

31 Might be a misspelling for Hayden.

32 November 1924, 12. Albert Comstock "was a ranch hand, who made the trip trailing cattle to Brooks/Red Deer place several times, good cowboy and cook" (Clay Chattaway). The other three Comstocks are not identified but appear to have been of Albert's family, possibly from the United States.

33 Voisey, *Vulcan,* 222–23.

34 Clay Chattaway.

35 Simon M. Evans, *The Bar U and Canadian Ranching History* (Calgary: University of Calgary Press, 2004), 109–48.

36 Cross got his start in the West in 1883 as a bookkeeper and veterinarian for the British American Company (Elofson, *Frontier Cattle Ranching,* 16–17).

37 Clay Chattaway.

38 Elofson, *So Far and Yet So Close,* 204–5.

39 "Range Notes," *Yellowstone Journal,* 15 October 1885.

40 Samuel Steele, *Forty Years in Canada: Reminiscences of the Great North-West, with some account of his service in South Africa*, ed. M. G. Niblett, 2nd ed. (Toronto: Prospero Books, 2000), 270-71.

41 Tom McKinnon, "The Stampede (or the Pinto Kid)," September 1924, 51-52; "Bill Patterson out of the chute at the Calgary Stampede," September 1924, 9: "Local News," February 1925, 10. See also, September 1924, 29.

42 Pp. 9, 11-14.

43 Pp. 11-12.

44 "Local News," April 1924, 3 and 6.

45 "Local News," 7.

46 "Local News," 1.

47 See note 46. Hendry is normally referred to as "Hendric" in the *Gazette*.

48 Voisey, *Vulcan*, 161-65.

49 Voisey, *Vulcan*, 164. "Chicago of the North" is our expression.

50 James Belich, *Replenishing the Earth: The Settler Revolution and the Rise of the Anglo World, 1783-1939* (Oxford: Oxford University Press, 2009), 324.

51 Elofson, *Frontier Cattle Ranching*, 63-80.

52 Frederick Ings, *Before the Fences: Tales from the Midway Ranch*, ed. J. Davis (Calgary: McAra Printing, 1980), 48.

53 Steele, *Forty Years in Canada*, 177.

54 On 10 May 1924, the Alberta Liquor Act was amended and prohibition ceased to exist in Alberta.

55 "Local News," November 1923, 1.

56 April 1924, 66.

57 "Selection from Mother Goose according to R.[obert] R.[aynor] – J.P.," March 1925, 59-60.

58 See also Elofson, *So Far and Yet So Close*, 70.

59 May 1924, 7.

60 See Russell, *Trails Plowed Under*, 159. "Is Gambling Prevalent Throughout the City," *Calgary Herald*, 17 October 1906.

61 See Elofson, *Frontier Cattle Ranching*, 103-4, 121-24; James Gray, *Red Lights on the Prairies* (Scarborough, ON: New American Library of Canada, 1973).

62 Char Smith, "Crossing the Line, American Prostitutes in Western Canada," *One Step Over the Line; Toward a History of Women in the North American Wests*, ed. Elizabeth Jameson and Sheila McManus (Edmonton: University of Alberta Press, 2008), 241-60.

63 The same was true on the mining frontier. For Butte Montana, see Mary Murphy, "Private Lives of Public Women," in *The Women's West*, ed. Susan Armitage and Elizabeth Jameson (Norman: University of Oklahoma Press, 1987), 191-205.

64 Frank W. Anderson, *Sheriffs and Outlaws of Western Canada* (Calgary: Frontier Publishing, n.d.), 48.

65 Hugh A. Dempsey, *The Golden Age of the Canadian Cowboy* (Calgary: Fifth House, 1995), 56.

66 "Claim City Is Run Wide Open," *Lethbridge Herald*, 9 July 1923. The *Herald* reported this sort of event quite often. See "Draws Stiff Fine," 29 December 1925; "To Get Out of City," 6 February 1925; "Blairmore Police Court," 24 March 1921.

67 "Local News," February 1925, 3. "The Empire Hotel was built in 1906 by William J. Stokes and Mr. Lewis at 118-9th Avenue SE, Calgary, Alberta. In 1920 the Empire was gutted by fire and after repairs was amalgamated with the Grand Central which at the time was owned by the Calgary Brewing and Malting Co. Ltd. The amalgamated hotel was named the Empire." In 1972, it was torn down to make way for the Calgary Convention Centre. See Archives Society of Alberta, online: https://albertaonrecord.ca/empire-hotel.

CHAPTER 10

1 *Preserving the Family Farm: Women, Community, and the Foundations of Agribusiness in the Midwest, 1900–1940* (Baltimore, MD: Johns Hopkins University Press, 1995), 17–70.

2 "Fire," 15–18.

3 High River had a fire department as early as 1912. Paul Voisey, *High River and the Times: An Alberta Community and Its Weekly Newspaper, 1905–1966* (Edmonton: University of Alberta Press, 2004), 27).

4 "In 1880 the National Bell Telephone Company had incorporated, through an Act of Parliament, the Bell Telephone Company of Canada. ... Which was thereby authorized to construct telephone lines over and along all public property and rights-of-way." There was general dissatisfaction with Bell for its failure to service less lucrative rural areas. Therefore, "in 1908 and 1909, Bell Telephone operations in Manitoba, Alberta and Saskatchewan were purchased by the provincial governments to be operated ... as provincially owned utilities." From that point, services were steadily extended to areas such as the Alberta foothills (*Canadian Encyclopedia*, http://www.thecanadianencyclopedia.ca/en/article/telephones/).

5 P. 16.

6 E. A. (Aubrey) Cartwright and John Thorpe (sometimes spelled Thorp) were the first to settle these properties around the turn of the twentieth century. Adjacent to the north of them was Macleay's "Half Way" place—halfway to the TL. All were on the east side of Highway 22.

7 Frank Scofield Sharpe, homesteaded SW 4-39-7-W5 in 1912 (Manitoba, Saskatchewan and Alberta, Canada, Homestead Grant Registers, 1872–1930).

8 A neighbouring farmer.

9   Clem Henson worked for Rocking P. He was a World War I veteran who came up from Texas with the Turkey Track ranch. Hugh Jenkins and Herb Thurber were teamsters who often hauled for Macleay.

10 The Martin family came to the Macleays' Bar S ranch in March 1925. Ostensibly Mr. Martin and, perhaps, Mrs. Martin worked on the ranch respectively as cow hand and domestic ("Local News," *Rocking P Gazette*, March 1925, 6).

11 Donald, Dunk, and Peter Comrie, World War I vets, homesteaded south of the Rocking P.

12 William and Jennie Gardiner settled on land to the west of the Rocking P in 1888.

13 Presumably NE 24-16-2-W5, which was a mile and a half southwest of the home place and thus about halfway between the home place and the TL ranch.

14 L. V. Kelly, *The Range Men*, 75th anniversary ed. (Calgary: Glenbow-Alberta Institute, 1988), 126. For other fires, see Elofson, *Cowboys, Gentlemen and Cattle Thieves; Ranching on the Western Frontier* (Montreal: McGill-Queen's University Press, 2000), 94–97.

15 For which see Bradford James Rennie, *The Rise of Agrarian Democracy: The United Farmers of Alberta, 1909–1921* (Toronto: University of Toronto Press, 2000), especially 138–60.

16 Glenbow Archives, Calgary, United Farmers of Alberta papers, Micro/ufa:*United Farmers of Alberta Annual Report and Year Book containing Reports of Officers and Committees for the year 1921 together with Official Minutes of the fourteenth Annual Convention, Calgary, January 17–21, 1922*, 40. For the politicizing of the UFA movement, see Rennie, *The Rise of Agrarian Democracy*, 179–206.

17 "Stereotype of Albertans as rednecks on social issues shattered: poll," https://ca.news.yahoo.com/blogs/dailybrew/stereotype-albertans-rednecks-social-issues-shattered-poll-232711906.html; Brenda Ward, "Rednecks, Rig Pigs, and Cowboys: Rural Masculinity in Albertan Country Music," http://ejournals.library.ualberta.ca/index.php/spacesbetween/article/view/19482: "This masculinity is heavily influenced by frontier and cowboy mythology. I will then show that country music acts as an expression of gender fantasy, recursively performing rural masculinity and thereby uncritically affirming and reinforcing the rural/redneck/cowboy categories. That is, country music appropriates frontier and cowboy mythology and acts as a vehicle of group values and ideologies, thereby forming and defining identity, including gender identity. I aim to demonstrate the interconnections between and the continuities of social practices, and the images that represent them. What ties the two together are narratives and ideals of cowboy and frontier mythology, often emerging from or depicting anxieties of a masculinity crisis."

18 Veronica Strong-Boag in "Pulling in Double Harness or Hauling a Double Load: Women, Work and Feminism on the Canadian Prairie," in *The Prairie West: Historical Readings*, ed. R. Douglas Francis and Howard Palmer, 2nd ed. (Edmonton: University of Alberta Press, 1992), 401–423, argues that women in the West had to fight exploitation by a patriarchal society. For historians' charges of racism see our next chapter.

19 September 1923, 31–34, written by "Carney Mulligan, Willow Bluff, Utah."

20 Pp. 33-34.

21 See Chapter 13.

22 Warren Elofson, *Frontier Cattle Ranching in the Land and Times of Charlie Russell* (Montreal: McGill Queen's University Press, 2004), 118-31.

23 The progression is visible in the population figures. Calgary and vicinity had 43,204 males and 37,214 females in 1916. In 1911 it had 39,657 males and 25,529 females. Macleod and vicinity had 19,379 males and 14,504 females in 1916. In 1911 it had 18,213 males and 12,548 females. Maple Creek and vicinity had 28,126 males and 19,424 females in 1916, and in 1911 it had 12,322 males and 7,408 females (Canada. *Census of Prairie Provinces*, 1916: Population and Agriculture, 44-127).

24 "Population of Canada, Provinces and Territories, 1921 to 2011, Topic-based tabulation: Age Groups (13) and Sex (3), http://www12.statcan.gc.ca/census-recensement/2011/dp-pd/tbt-tt/Rp-eng.cfm?LANG=E&APATH=3&DETAIL=0&DIM=0&FL=A&FREE=0&GC=0&GID=0&GK=0&GRP=0&PID=102186&PRID=0&PTYPE=101955&S=0&SHOWALL=0&SUB=0&Temporal=2011&THEME=88&VID=0&VNAMEE=&VNAMEF=

25 Edward C. Abbott and Helen Huntington Smith, *We Pointed Them North: Recollections of a Cowpuncher*, 2nd ed. (Norman: University of Oklahoma Press, 1955), 188-89.

26 Con Price, *Memories of Old Montana* (Pasadena, CA: Trail's End Publishing, 1945), 37.

27 "Local News," 6-7.

28 Charles Dew, a nearby rancher.

29 R. Raynor, "Local News," January 1925, 54.

30 Voisey, *Vulcan: The Making of a Prairie Community* (Toronto: University of Toronto Press, 1982), 221-46.

31 Frederick Ings, *Before the Fences: Tales from the Midway Ranch*, ed. J. Davis (Calgary: McAra Printing, 1980), 76.

32 "Local News," February 1925, 4.

33 March 1925, 29-33.

34 Pp. 12-17.

35 "The Dying Cowboy," September 1924, 43-49.

36 "Personal," December 1923, 55-58.

37 Sandra Meyres, *Westering Women and the Frontier Experience, 1800-1915* (Albuquerque: University of New Mexico Press, 1982), 165.

38 *A Female Economy: Women's Work in a Prairie Province* (Montreal: McGill University Press, 1998), 95.

39 Rennie, *The Rise of Agrarian Democracy*, 114-16.

40 Murphy, one of the "famous five" who fought the "persons case" for the legal recognition of women as persons, which they won in 1929, author Janey Canuck articles and books, executive of Canadian Women's Press Club, leader in campaign for the 1917 Dower Act in Alberta; Edwards, "persons case," fought for Alberta Dower Act, author,

Legal Status of Women; McClung, "persons case," editor *Manitoba Monthly*, founder Women's Christian Temperance Union, MLA Alberta 1921–26, author *Sowing Seeds in Danny, Stream Runs Fast* etc.; Parlby, "persons case," president United Farm Women of Alberta, MLA and minister without portfolio, delegate to the League of Nations; McKinney, "persons case," Dower Act, as MLA for Non-Partisan League in 1917 one of the first women elected to a legislative assembly in the British Empire; McNaughton, president of the Saskatchewan Women's Grain Growers Association, president of the Interprovincial Council of Farm Women; McNeal, president of the SWGGA; McCallum, women's editor of the *Country Guide*; Binnie-Clark, author *A Summer on the Canadian Prairie* and *Wheat and Women*; Gale, alderwoman Calgary and as such first woman elected to any government position in the British Empire (1917), trustee Calgary Board of Education.

41   Women's suffrage was granted 28 January 1916 in Manitoba, 14 May 1916 in Saskatchewan, 19 May 1916 in Alberta, 5 April 1917 in BC, and 24 May 1918 federally.

42   Some scholars would argue that women like the above took the leadership in the fight against the victimization and exploitation of their frontier sisters who were subjected to the hardships and privations of a very difficult life. See, for instance, Strong-Boag, "Pulling in Double Harness or Hauling a Double Load."

CHAPTER 11

1   *Patterns of Racism: Attitudes Toward Chinese and Japanese in Alberta* (Ottawa: University of Ottawa Press, 1979); also by Palmer, *Patterns of Prejudice: A History of Nativism in Alberta* (Toronto: McClelland and Stewart, 1982); "Strangers and Stereotypes: The Rise of Nativism in Alberta, 1880–1920," in *The Prairie West: Historical Readings*, ed. R. Douglas Francis and Howard Palmer, 2nd ed. (Edmonton: University of Alberta Press, 1992), 308–34; "Reluctant Hosts: Anglo-Canadian Views of Multiculturalism in the Twentieth Century," in *Readings in Canadian History: Post-Confederation*, ed. R. Douglas Francis and Donald B. Smith, 7th ed. (Toronto: Thomson Nelson Learning, 2006), 143–61.

2   Hugh A. Dempsey, "Cypress Hills Massacre," *Montana Magazine* 3, no. 4 (Autumn 1953): 1–9.

3   *Great Plains Quarterly* 24 (Spring 2004): 96. The police role in supporting government, big business, and the cattlemen against Indigenous people and immigrant workers is explored in Graybill's *Policing the Great Plains: Rangers, Mounties and the North American Frontier, 1875–1910* (Lincoln: University of Nebraska Press, 2007). In the Mounties' defence, we would mention that they were known at times to side with the Indigenous peoples against corporations like the Canadian Pacific Railway; see "In Town and Out," *Macleod Gazette*, 1 July 1882.

4   *Clearing the Plains: Disease, Politics of Starvation, and the Loss of Aboriginal Life* (Regina: University of Regina Press, 2013).

5   Paul Voisey, *Vulcan: The Making of a Prairie Community* (Toronto: University of Toronto Press, 1982), 221–46.

6 Abram de Swaan, *Killing Compartments: The Mentality of Mass Murder* (New Haven: Yale University Press, 2015).

7 Thus the lack of religious or racial conflict in many multicultural rural western societies; see Voisey, *Vulcan*, 175–98, 221–46.

8 Glenbow Archives, Calgary, M376: Mrs. Charles Inderwick, Diary and Personal Letters from the North Fork Ranch, "Letter written in the Canadian North West Territory in 1884, the East Range Ranche, May 13th 1884."

9 Voisey, *Vulcan*, 174–98

10 P. 9.

11 For instance, "Jokes," November 1924, 87; "Bills Last Romance," January 1925, 52; "All's Well That Ends Well," April 1925, 55. On the numerous instances in which a church is mentioned in any of the published stories, it invariably is in connection with a social service it provided, in a particular marriage, rather than worship.

12 December 1924, 44.

13 L. V. Kelly, *The Range Men*, 75th anniversary ed. (Calgary: Glenbow-Alberta Institute, , 1988), 60–62.

14 Glenbow Archives, Macleod papers, M776-14a: Macleod to Mary Macleod, 3 June 1880.

15 Captain Richard Burton Deane, *Mounted Police Life in Canada* (Toronto: Prospero, 2001), 149.

16 Ings, *Before the Fences: Tales from the Midway Ranch,* ed. by J. Davis (Calgary: McAra Printing, 1980) 34.

17 "During the first few censuses after Confederation, the British Isles were the main source of immigration, accounting for 83.6% of the foreign-born population in the 1871 Census, or close to half a million people. Immigrants from the United States (10.9%), Germany (4.1%) and France (0.5%) were far behind. The population of immigrants born in European countries other than those of the British Isles started to increase in the late 1800s, slowly at first and then more rapidly, peaking in the 1970s. This transformation consisted of three major waves. The first wave began in the late 1800s and early 1900s, with the arrival of new groups of immigrants from Eastern Europe (Russians, Polish and Ukrainians), Western Europe and Scandinavia, ("150 years of immigration in Canada," 29 June 2016, http://www.statcan.gc.ca/pub/11-630-x/11-630-x2016006-eng.htm).

18 Warren Elofson, *Somebody Else's Money: The Walrond Ranch Story, 1883–1907* (Calgary: University of Calgary Press, 2009), 161–62.

19 P. 8.

20 "Celebrate Purchase of Eden Ranch," *Lethbridge Herald*, 28 October 1948.

21 "Results of Last Month's Competition," May 1924, 38.

22 "Competition," November 1924, 44.

23 "Grand Competition," December 1923, 36.

24 "This Month's Competition, Count the Dots in the Circle," October 1924, 50.

25 "Thanks," February 1925, 34.

26 "Jokes," November 1924, 88.

27 "Jokes," March 1924, 69.

28 By Ethel Watts, November 1924, 39–43.

29 P. 43.

30 P. 92. The piece is credited to Annabella Trunk of Tunnerville, Ontario. In two other citations in the *Gazette* Trunk is cited as Annabelle. Her work seems to be copied from another print media source. The Longfellow poem reads:

> Thus departed Hiawatha,
>
> Hiawatha the Beloved,
>
> In the glory of the sunset,
>
> In the purple mists of evening,
>
> To the regions of the home-wind,
>
> Of the Northwest wind Keewajdin,
>
> To the Islands of the Blessed,
>
> To the kingdom of Ponemah,
>
> To the land of the Hereafter!

31 Voisey, *Vulcan*, 214.

32 Palmer, "Strangers and Stereotypes," 316.

33 Pp. 36–37.

34 I am indebted to Dr. David Wright for the translations in the following six notes. *Translator's note*: "These two phrases are famous in the modern Chinese language and trace back to a memorial (a written communication) by Sun Yat-sen (widely regarded as the father of modern China) in 1894 to Li Hongzhang, a high government official of the Manchu Qing dynasty then ruling China. Li Hongzhang ignored Sun Yat-sen because Sun had no academic degrees in traditional Confucian learning, and after this contemptuous dismissal and the victory of Japan over China in a war between the two countries in 1895, Sun Yat-sen gave up on trying to reform the Manchu Qing dynasty and became a revolutionary whose forces overthrew it in 1911 and founded the Republic of China. The meaning of the two phrases here is that the people are the basis of the state, and in turn that the basic need of the people is for food (and thus a reference to agricultural production in Canada). The extended meanings are that food is more important to the people than the state and that a country cannot be strong and prosperous if its people are not well nourished."

35 *Translator's note*: "In other words, both men and women."

36 *Translator's note*: "Maishen here seems to be a place name, but the location is unclear."

37 *Translator's note*: "i.e., the morning. The meaning here is that Chinese daffodils are at their most beautiful in the morning."

38 *Translator's note*: "i.e., Chinese daffodils."

39 *Translator's note*: "Maiwei is a place name in Guangdong province, approximately 80 kilometres NNW of Guangzhou."

40 *Translator's note*: "This is a place name, but the Chinese characters are illegible. The writer is quite educated and literate and uses the old classical or literary style of Chinese, not modern colloquial Chinese. He also uses no punctuation, which is indicative of his style dating to before 1917, if not the actual date of the composition. At the end of the small piece the author gives his name and where he was living or staying at the time he wrote this.

The writing here is not concentrated on any one thing. It is what the Chinese call 'random jottings' written into little notebooks that literate and educated people took along with them to jot down their thoughts or impressions on the spot, before they forgot them. (Think Moleskines, I guess.) The random jottings here seem to pertain to three things: first, a statement that Chinese daffodils are the most beautiful of all vistas; second, a small comment on a famous statement by Sun Yat-sen, who said that the people were the basis of the state and that food was the basic need of the people; and third, a few sentences more or less arguing that Chinese daffodils are the greatest of all flowers, greater even than peonies. In Chinese culture, daffodils are very appropriate flowers to give to sick people. The daffodil symbolizes good fortune in the Chinese culture. In fact, it is so esteemed for its ability to bring forth positive things that it is the official symbol of the Chinese New Year."

41 May 1924, 67; quote, 74–75.

42 James Morrow Walsh. In 1875, Walsh was sent to the Cypress Hills in command of B Division to establish an independent post (Fort Walsh), which he was allowed to name for himself.

43 Macleod papers, M776-14a: Macleod to Mary Macleod, 29 July 1878.

44 *Westering Women and the Frontier Experience, 1800–1915* (Albuquerque: University of New Mexico Press, 1982), 96–97.

45 P. 18.

46 As we have seen, Colonel Macleod used the term.

47 Edward C. Abbott and H. Huntington Smith, *We Pointed Them North: Recollections of a Cowpuncher*, 2nd ed. (Norman: University of Oklahoma Press, 1955), 222–33; E. J. "Bud" Cotton,, *Buffalo Bud: Adventures of a Cowboy* (North Vancouver: Hancock House, 1981), 148.

48 *We Pointed Them North*, 149.

49 *We Pointed Them North*, 208–9.

50 By "Coyote Cal," November 1923, 33–36.

51 Brian Dippie, ed., *Charlie Russell Roundup: Essays on America's Favorite Cowboy Artist* (Helena: Montana Historical Society, 1999), 146.

52 Grant McEwan, *John Ware's Cow Country*, 3rd ed. (Vancouver: Greystone Books, 1995). John Ware Ridge (formerly Nigger John Ridge), Mount Ware, and Ware Creek, all near the Ware ranch, are named after him and the family.

53 Quoted in David Breen, "John Ware," *Dictionary of Canadian Biography*, http://www.biographi.ca/en/bio/ware_john_13E.html.

54 Glenbow Archives, Calgary, M-1281-2: Slim Marsden, "Reminiscences; John Ware, Famous Cowboy, of the Bar U," n.d., http://www.glenbow.org/collections/search/findingAids/archhtm/extras/ware/m-1281-2.pdf.

55 Marsden called Ware "the Whitest Man in the North West Territories" (Reminiscences). See also Ings, *Before the Fences*L. V. Kelly, *The Range Men*, 75th anniversary ed. (Calgary: Glenbow-Alberta Institute, 1988), 5

56 Kelly, *The Range Men*, 5.

57 P. 76.

58 May 1924, 93.

59 Surnames such as McKinnon, Hendrie, and McDonal are, like Macleay, Scottish.

60 "Jokes," November 1924, 89.

61 "The Passing of an Old Cowpuncher,"13

CHAPTER 12

1 "Flivver" is slang for decrepit old car.

2 "Local News," January 1924, 3; cartoon drawing, p. 13.

3 For instance, "The Romantic Hour," December 1924, 17.

4 "Ads," December 1923, 62; "Adds" [sic], September 23, 48; "Romantic Hour," December 1924, 22; "Ads," November 1923, 65; "Ads," March 1924, 73.

5 Larger items were usually delivered to a depot or post office, but anyone in the family, or even a neighbour, could retrieve them when picking up necessities.

6 Georgina Helen Thompson, *Crocus and Meadowlark Country: Recollections of a Happy Childhood and Youth in Southern Alberta* (Edmonton: Institute of Applied Art, 1963), 83, quoted in Paul Voisey, *Vulcan: The Making of a Prairie Community* (Toronto: University of Toronto Press, 1988), 24.

7 See also May 1924, 100; "Ads," September 1924, 79; "High River Hats," January 1925, 12; "Hat Styles for cow girls," April 1925, 46. For men's fashions, see March 1925, 74.

8 His numerous trips to Calgary and other towns are recorded in the *Rocking P Gazette*. For instance, "Local News," February 1925, 8: "Mr. R. Macleay has spent the past week in Calgary attending all the meetings etc."; "Local News," January 1924, 3: "Mr. Macleay has been to town and back several times this month and has had to shovel several times also."

9 See *Blairmore Enterprise*.

10 See *Macleod Times*.

11 See *Empress Express*.

12 See *Macleod Gazette*.

13 Ted Nelson was principally a horse wrangler. Jesse Walters lived at the cow camp on the Bar S.

14 February 1925, 5–8.

15 "Notice," January 1925, 22. Frank Van Eden on the Bar S.

16 For instance, Val, Jim, Tex, and Bill are all found in "Shooting Through Life" (13–19), and Tex, Bill, Stewart, and Tom are in "Shorty Passes On" (35–44) in the January 1925 issue.

17 December 1923, 21.

18 April 1924, 19–20.

19 Sancho and Sawndy were dogs. The latter was experiencing the infirmities of old age.

20 January 1925, 45–47.

CHAPTER 13

1 Warren Elofson, "Other People's Money: Patrick Burns and the Beef Plutocracy," *Prairie Forum* 32, no. 2 (Summer 2007): 235–36.

2 Paul Voisey, *Vulcan: The Making of a Prairie Community* (Toronto: University of Toronto Press, 1982), 310. Voisey cites the following works: Morton Rothstein, "The Big Farm: Abundance and Scale in American Agriculture," *Agricultural History* 49, no. 4 (October 1975): 585; Paul Wallace Gates, "Large-Scale Farming in Illinois, 1850 to 1870," *Agricultural History* 6, no. 1 (January 1932): 14–25; Harold E. Briggs, "Early Bonanza Farming in the Red River Valley of the North," *Agricultural History*, 6, no. 1 (January 1932): 26–37; Stanley Norman Murray, *The Valley Comes of Age: A History of Agriculture in the Valley of the Red River of North, 1812-1920* (Fargo: North Dakota Institute for Regional Studies, 1967): 131–38; Hiram M. Drache, *The Day of the Bonanza: A History of Bonanza Farming in the Red River Valley of the North* (Fargo: North Dakota Institute for Regional Studies, 1964), and "Bonanza Farming in the Red River Valley," *Historical and Scientific Society of Manitoba Transactions* 3, no. 24 (1967–68): 53–64; E. C. Morgan, "The Bell Farm," *Saskatchewan History* 19, no. 2 (Spring 1966): 41–60; Don G. McGowan, *Grassland Settlers: The Swift Current Region during the Early Years of the Ranching Frontier* (Regina: Canadian Plains Research Center, 1976), 57–59; and Grant McEwan, *Illustrated History of Western Canadian Agriculture* (Regina: Western Producer Prairie Books, 1980), 57–79.

3 "One Earth Farms Restructures," https://www.producer.com/2014/05/one-earth-farms-restructures/, Mike Beretta, company chief executive officer, 15 May 2014,

4 Glenbow Archives, New Walrond Ranche papers, count books, M8688-37.

5 Figures are also given from time to time in the company letters and annual reports. Thus, for instance, on 21 October 1905 McEachran wrote to A. M. Walrond informing

him that the ranch had branded nearly 2,300 head (New Walrond Ranche papers, M8688-8); see also M8688-2: "Sixth Annual Report of the New Walrond Ranche Company Limited," for the year ended 31 December 1903.

6   Warren Elofson, *Somebody Else's Money: The Walrond Ranch Story, 1883-1907* (Calgary: University of Calgary Press, 2009), 191-220.

7   Clay Chattaway.

8   Elofson, *Somebody Else's Money*, 141-52.

9   Frederick Ings, *Before the Fences: Tales from the Midway Ranch*, ed. J. Davis (Calgary: McAra Printing, 1980), 78.

10 Simon M. Evans, *The Bar U and Canadian Ranching History* (Calgary: University of Calgary Press, 2004), 163.

11 Ibid., 149-70.

12 Glenbow Archives, A.E. Cross papers, M8780-112: Cross to A.R. Springett, 10 November 1902.

13 In the period from 1924 to 1928 Macleay ran horses with his cattle on the 76 ranch. And in March 1925 he actually had horses on feed ("Local News," *Rocking P Gazette*, 4).

14 Clay Chattaway, "Ranching Changes 1945 to 2005," Macleay family papers.

15 "Number and Average Size of Alberta Farms, 1961-2006," http://www1.agric.gov.ab.ca/$department/deptdocs.nsf/all/sdd12892/$FILE/figure35.pdf.

16 *Census of Canada*, 1971, Agriculture Alberta, "Large farms and census farms by province."

17 See http://www.guy-sports.com/months/jokes_farming.htm. We substituted "Alberta" for Montana and changed the government agency to Alberta government.

18 Clay Chattaway, "Ranching Changes 1945 to 2005," Macleay family papers.

19 It has very recently also been deposited in hard copy, in the Glenbow Alberta Library and Archives in Calgary.

BIBLIOGRAPHY

I. Manuscript Collections

Bar S Ranch, Nanton, Alberta. Macleay family papers.

Bar S Ranch. Clay Chattaway's Notes, 30 October 2013. "Chattaway Section 2 cleaning Oct 30."

Canada. *Sessional Papers*. North West Mounted Police Annual Reports.

Glenbow Archives, Calgary. Canadian Agricultural, Coal and Colonization Company papers, Stair Ranch Letterbook.

———. Burns papers, M160.

———. Billy and Evelyn Cochrane papers.

———. A. E. Cross papers.

———. Herbert M. Hatfield papers.

———. Mrs. Charles Inderwick, Diary and Personal Letters from the North Fork Ranch.

———. Macleod papers, M776-14a.

———. New Walrond Ranche papers.

———. United Farmers of Alberta papers, Micro/ufa.

Montana Historical Society Library and Archives, Helena. Power papers.

Provincial Archives of Alberta, Edmonton. 72,27/SE: Violet LaGrandeur, "Memoirs of a Cowboy's Wife."

II. Newspapers and Pamphlets

Farm and Ranch Review

Lethbridge Herald

Macleod Gazette

Medicine Hat News

Pincher Creek Echo

Regina Leader

Rocky Mountain Husbandman

III. Government Reports

United Kingdom. "Agricultural Interests Commission, Reports of the Assistant Commissioners presented to both Houses of Parliament by Command of Her Majesty." *British Parliamentary Papers, Area Studies: United States of America, I. Agriculture, 1878–99.*

United States. *Range and Ranch Cattle Traffic of the United States.* New York: Office of Poor, White and Greenough, 1885.

IV. Books and Articles

Abbott, Edward C., and H. Huntington Smith. *We Pointed Them North: Recollections of a Cowpuncher,* 2nd ed. Norman: University of Oklahoma Press, 1955.

Abbott, J. S. C., *Christopher Carson familiarly known as Kit Carson.* New York: Dodd, Mead and Co., 1874.

Adams, B. W., et al. *Range Plant Communities and Range Health Assessment Guidelines for the Foothills Fescue Natural Subregion of Alberta, Foothills Fescue Range Plant Community Guide, Alberta.* Lethbridge: Alberta Sustainable Resource Development, 2005.

Anderson, Frank W. *Sheriffs and Outlaws of Western Canada.* Calgary: Frontier Publishing, n.d.

Attwood, Bain, and S. G. Fostern, eds. *Frontier Conflict.* Canberra: National Museum of Australia, 2003.

Belich, James. *Replenishing the Earth: The Settler Revolution and the Rise of the Anglo World, 1783–1939.* Oxford: Oxford University Press, 2009.

Bowers, William L. *The Country Life Movement in America, 1900–1920.* Port Washington, NY: Kennikat Press, 1974.

Breen, David H. *The Canadian Prairie West and the Ranching Frontier, 1874–1924.* Toronto: University of Toronto Press, 1983.

Brégent-Heald, Dominique. *Borderland Films: American Cinema, Mexico, and Canada during the Progressive Era.* Lincoln: University of Nebraska Press, 2015.

Brisbin, James S. *The Beef Bonanza or How to get Rich on the Plains; being a description of cattle-growing, sheep-farming, horse-raising, and dairying in the West.* Philadelphia: J.B. Lippincott & Co., 1881.

Canada. *Farming and Ranching in Western Canada.* [Montreal, 1890].

Carter, Sarah. "'He Country in Pants' No Longer—Diversifying Ranching History." In *Cowboys, Ranchers and the Cattle Business: Cross Border Perspectives on Ranching History,* edited by Simon Evans, Sarah Carter, and Bill Yeo, 155–66. Calgary: University of Calgary Press, 2000.

Chattaway, Clay, and Warren Elofson. *The Rocking P Gazette and Western Canadian Ranching History.* University of Calgary, Libraries and Cultural Resources Digital Collections, 2014, online: http://contentdm.ucalgary.ca/digital/collection/rpg.

Connor, Ralph. *Sky Pilot: A Tale of the Foothills.* Chicago: R. H. Revell, 1899.

Cotton, E. J. "Bud." *Buffalo Bud: Adventures of a Cowboy.* North Vancouver: Hancock House, 1981.

Cox, James. *Historical and Biographical Record of the Cattle Industry and the Cattlemen of Texas and Adjacent Territory.* Saint Louis: Woodward and Tiernan Printing Co., 1895.

Cross, Alfred Ernest. "The Roundup of 1887." *Alberta Historical Review* 13, no. 2 (Spring 1965): 23–27.

Cunfer, Geoff. *On the Great Plains: Agriculture and the Environment.* College Station: Texas A&M University Press, 2005.

Dale, Edward Everett. *The Range Cattle Industry: Ranching on the Great Plains from 1865 to 1925.* Norman: University of Oklahoma Press, 1960.

Daschuk, James. *Clearing the Plains: Disease, Politics of Starvation and the Loss of Aboriginal Life.* Regina: University of Regina Press, 2013.

Deane, Captain Richard Burton. *Mounted Police Life in Canada.* Toronto: Prospero, 2001.

Dempsey, Hugh A. "Cypress Hills Massacre." *Montana Magazine* 3, no. 4 (Autumn 1953): 1–9.

———. The Golden Age of the Canadian Cowboy. Calgary: Fifth House, 1995.

de Swaan, Abram. *Killing Compartments: The Mentality of Mass Murder.* New Haven, CT: Yale University Press, 2015.

Dippie, Brian, ed. *Charlie Russell Roundup: Essays on America's Favorite Cowboy Artist.* Helena: Montana Historical Society, 1999.

Elofson, Warren M. *Cowboys, Gentlemen and Cattle Thieves: Ranching on the Western Frontier.* Montreal: McGill-Queen's University Press, 2000.

———. *Frontier Cattle Ranching in the Land and Times of Charlie Russell.* Montreal: McGill-Queen's University Press, 2004.

———. "Grasslands Management in Southern Alberta: The Frontier Legacy." *Agricultural History,* 86, no. 4 (2012): 143–68.

———. "Not Just a Cowboy: The Practice of Ranching in Southern Alberta, 1881–1914." *Canadian Papers in Rural History* 10 (1996): 205–16.

———. "Other People's Money: Patrick Burns and the Beef Plutocracy." *Prairie Forum* 32, no. 2 (Summer 2007): 235–36.

———. *So Far and Yet So Close: Frontier Cattle Ranching in Western Prairie Canada and the Northern Territory of Australia.* Calgary: University of Calgary Press, 2015.

———. *Somebody Else's Money: The Walrond Ranch Story, 1883–1907.* Calgary: University of Calgary Press, 2009.

Evans, Simon M. "Stocking the Canadian Ranges." *Alberta History* 26, no. 1 (Summer 1978): 1-8.

———. "Tenderfoot to Rider: Learning 'Cowboying' on the Canadian Ranching Frontier During the 1880s." In *Cowboys, Ranchers and the Cattle Business: Cross Border Perspectives on Ranching History*, edited by Simon Evans, Sarah Carter, and Bill Yeo, 61-80. Calgary: University of Calgary Press, 2000.

———. *The Bar U and Canadian Ranching History*. Calgary: University of Calgary Press, 2004.

Feldberg, John, and Warren Elofson. "Financing the Palliser Triangle, 1908-1913." *Great Plains Quarterly* 18 (Summer 1998): 257-68.

Fitch, L. B. Adams, P. Ag, and K. Oshaughnessy. *Caring for the Green Zone: Riparian Areas and Grazing Management*. 3rd ed. Available online: http://www.cowsandfish.org/riparian/caring.html.

Fletcher, Robert S. "The End of the Open Range in Eastern Montana." *Mississippi Valley Historical Review* 16 (September 1929): 188-211.

Foothills Historical Society. *Chaps and Chinooks: A History West of Calgary*. Calgary: Foothills Historical Society, 1976.

Foran, Max. *Trails and Trials: Markets and Land Use in the Alberta Beef Cattle Industry, 1881-1918*. Calgary: University of Calgary Press, 2003.

Fowke, Edith. "American Cowboy and Western Pioneer Songs in Canada." *Western Folklore* 21 (1962): 247-56.

Garceau-Hagen, Dee. *The Important Things of Life: Women, Work, and Family in Sweetwater County, Wyoming, 1880-1929*. Lincoln: University of Nebraska Press, 1997.

Gray, James. *Men Against the Desert*. Saskatoon: Western Producer Prairie Books, 1967.

———. *Red Lights on the Prairies*. Scarborough: The New American Library of Canada, 1973.

Graybill, Andrew. "Rangers, Mounties, and the Subjugation of Indigenous Peoples, 1870-1885." *Great Plains Quarterly* 24 (Spring 2004): 83-100.

———. *Policing the Great Plains, Rangers, Mounties and the North American Frontier, 1875-1910*. Lincoln: University of Nebraska Press, 2007.

Hall, S. S. *Stampede Steve; or, the Doom of the Double Face*. New York: Beadle and Adams, 1884.

Herbert, Rachel. *Ranching Women in Southern Alberta*. Calgary: University of Calgary Press, 2017.

Hopkins, Monica. *Letters from a Lady Rancher*. Halifax: Formac Publishing, 1983.

Ingraham, Prentiss. *Buffalo Bill, from Boyhood to Manhood; Deeds of Daring, Scenes of Thrilling Peril, and Romantic Incidents in the Early Life of W.F. Cody, the Monarch of the Borderland*. New York: Beadle and Adams, [1882].

Ings, Frederick. *Before the Fences: Tales from the Midway Ranch*. Edited by J. Davis. Calgary: McAra Printing, 1980.

Jameson, Shelagh S. "Women in the Southern Alberta Ranch Community, 1881–1914." In *The Canadian West: Social Change and Economic Development*, edited by Henry C. Klassen, 63–78. Calgary: University of Calgary Comprint Publishing, 1977.

Jennings, John. *The Cowboy Legend: Owen Wister's Virginian and the Canadian-American Frontier*. Calgary: University of Calgary Press, 2015.

Jones, David C. "An Exceedingly Risky and Unremunerative Partnership: Farmers and the Financial Interests Amid the Collapse of Southern Alberta." In *Building Beyond the Homestead*, edited by David C. Jones and Ian MacPherson, 207–27. Calgary: University of Calgary Press, 1985.

———. "'There Is Some Power About the Land': The Western Agrarian Press and Country Life Ideology." In *The Prairie West: Historical Readings*, 2nd ed., edited by R. Douglas Francis and Howard Palmer, 455–74. Edmonton: University of Alberta Press, 1992.

Jordan, Terry G. *North American Cattle Ranching Frontiers: Origins, Diffusion and Differentiation*. Albuquerque: University of New Mexico Press, 1993.

Kelly, Leroy V. *The Range Men*. 75th anniversary ed. Calgary: Glenbow-Alberta Institute, 1988.

Kinnear, Mary. *A Female Economy: Women's Work in a Prairie Province*. Montreal: McGill-Queen's University Press, 1998.

Klassen, Henry C. "A Century of Ranching at the Rocking P and Bar S." In *Cowboys, Ranchers and the Cattle Business: Cross Border Perspectives on Ranching History*, edited by Simon Evans, Sarah Carter, and Bill Yeo, 101–22. Calgary: University of Calgary Press, 1997.

Klein, Kerwin Lee. "Reclaiming the 'F' Word, or Being and Becoming Postwestern." *Pacific Historical Review* 65, no. 2 (May 1996): 179–215.

Knupp, Lillian. *Leaves from the Medicine Tree: A history of the area influenced by the tree, and biographies of pioneers and oldtimers who came under its spell prior to 1900*. Lethbridge: High River Pioneers' and Old Timers Association, 1960.

Latham, Hiram. *Trans-Missouri Stock Raising: The Pasture Lands of North America; Winter Grazing*. Omaha, NB: Daily Herald Steam Printing House, 1871.

Lavington, Harold "Dude." *Nine Lives of a Cowboy*. Victoria: Sono Nis Press, 1982.

Lawrence, H. F. "Early Days in the Chinook Belt." *Alberta Historical Review* 13, no. 1 (Winter 1965): 9–19.

Limerick, Patricia Nelson. "Going West and Ending Up Global." *Western Historical Quarterly* 32, no. 1 (Spring 2001): 4–23.

———. *The Legacy of Conquest: The Unbroken Past of the American West*. New York: Norton, 1987.

———. "Turnerians All: The Dream of a Helpful History in an Intelligible World." *American Historical Review* 100, no. 3 (June 1995): 697–716.

Macdonald, James. *Food from the Far West*. New York: Orange Judd, 1878.

MacEwan, Grant. *John Ware's Cow Country*. 3rd ed. Vancouver: Greystone Books, 1995.

Magnan, André. *When Wheat Was King: The Rise and Fall of the Canada-UK Wheat Trade*. Vancouver: UBC Press, 2016.

McCoy, Joseph. *Historic Sketches of the Cattle Trade of the West and Southwest*. Kansas City, MO: Ramsay, Millett and Hudson, 1874.

McCullough, A. B. "Winnipeg Ranchers: Gordon, Ironside and Fares." *Manitoba History* 41 (Spring/Summer 2001): 18–25.

McEachran, Duncan. *A Journey Over the Plains: From Fort Benton to Bow River and Back*. [Montreal: 1881].

McGerr, Michael. *A Fierce Discontent: The Rise and Fall of the Progressive Movement in America, 1870–1920*. New York: Free Press, 2003.

McIntyre, Stuart. *A Concise History of Australia*. Cambridge: Cambridge University Press, 1999.

McIntyre, William H. *A Brief History of the McIntyre Ranch*. [1948].

McPherson, Kathryn. "Was the Frontier Good for Women? Historical Approaches to Women and Agricultural Settlement in the Prairie West, 1870–1925." *Atlantis: A Women's Study Journal* 25, no. 1 (2000): 75–86.

Merchant, Carolyn. *Reinventing Eden: The Fate of Nature in Western Culture*. New York: Routledge, 2003.

Merrill, A., and J. Jacobson. *Montana Almanac*. Helena: Falcon Books, 1997.

Meyres, Sandra. *Westering Women and the Frontier Experience, 1800–1915*. Albuquerque: University of New Mexico Press, 1982.

Murphy, Mary. "Private Lives of Public Women." In *The Women's West*, edited by Susan Armitage and Elizabeth Jameson, 191–205. Norman: University of Oklahoma Press, 1987.

Nelson, J. G. "Some Reflections on Man's Impact on the Landscape of the Canadian Prairies and Nearby Areas." In *The Prairie Provinces*, edited by P. J. Smith. Toronto: University of Toronto Press, 1972.

Neth, Mary. *Preserving the Family Farm: Women, Community, and the Foundations of Agribusiness in the Midwest, 1900–1940*. Baltimore: Johns Hopkins University Press, 1995.

Nettelbeck, Amanda, and Robert Foster. *In the Name of the Law: William Willshire and the Policing of the Australian Frontier*. Kent Town, SA: Wakefield Press, 2007.

Osgood, Ernest Staples. *The Day of the Cattleman*. Minneapolis: University of Minnesota Press, 1929.

Palmer, Howard. *Patterns of Prejudice: A History of Nativism in Alberta*. Toronto: McClelland and Stewart, 1982.

———. *Patterns of Racism: Attitudes Toward Chinese and Japanese in Alberta*. Ottawa: University of Ottawa Press, 1979.

———. "Reluctant Hosts: Anglo-Canadian Views of Multiculturalism in the Twentieth Century." In *Readings in Canadian History: Post-Confederation,* edited by R. Douglas Francis and Donald B. Smith, 7th ed., 143–61. Toronto: Thomson Nelson Learning, 2006.

———. "Strangers and Stereotypes: The Rise of Nativism in Alberta, 1880–1920." In *The Prairie West: Historical Readings,* edited by R. Douglas Francis and Howard Palmer, 2nd ed., 308–34. Edmonton: University of Alberta Press, 1992.

Patterson, Paul E., and Joy Poole. *Great Plains Cattle Empire: Thatcher Brothers and Associates, 1875–1945.* Lubbock: Texas Tech University Press, 2000.

Pincher Creek Historical Society. *Prairie Grass to Mountain Pass.* Pincher Creek, AB, 1974.

Price, Con. *Memories of Old Montana.* Pasadena, CA: Trail's End Publishing, 1945.

Rasmussen, Linda, et al., eds. *A Harvest Yet to Reap: A History of Prairie Women.* Toronto: The Women's Press, 1976.

Rathborne, St George Henry. *Sunset Ranch.* London: Shumen Sibthorp, [1902].

Rennie, Bradford James. *The Rise of Agrarian Democracy: The United Farmers and Farm Women of Alberta, 1909–1921.* Toronto: University of Toronto Press, 2000.

Riley, Harold W. "Herbert William (Herb) Millar, Pioneer Rancher." *Canadian Cattlemen* 4, no. 4 (March 1942).

Roen, Hazel Bessie. *The Grass Roots of Dorothy, 1895–1970,* 2nd ed. Calgary: Northwest Printing and Lithographing, 1971.

Roosevelt, Theodore R. *Ranch Life and the Hunting Trail.* London: T. Fisher Unwin, [1888].

Russell, Charles. *Trails Plowed Under: Stories of the Old West.* Introduction by W. Rogers and B. W. Dippie. Lincoln: University of Nebraska Press, 1996.

Sharp, Paul. "Three Frontiers: Some Comparative Studies of Canadian, American, and Australian Settlement." *Pacific Historical Review* 24, no. 4 (November 1955): 369–77.

Smith, Char. "Crossing the Line, American Prostitutes in Western Canada." In *One Step Over the Line: Toward a History of Women in the North American Wests,* edited by Elizabeth Jameson and Sheila McManus, 241–260. Edmonton: University of Alberta Press, 2008.

Stanley, George F. G. "Western Canada and the Frontier Thesis." *Canadian Historical Association Annual Report* (1940), 105–14.

Steele, Samuel. *Forty Years in Canada: Reminiscences of the great north-west with some account of his service in South Africa,* edited by M. G. Niblett. 2nd ed. Toronto: Prospero Books, 2000.

Stegner, Wallace. *Wolf Willow: A History, a Story, and a Memory of the Last Plains Frontier.* New York: Penguin Books, 1955.

Strong-Boag, Veronica. "Pulling in Double Harness or Hauling a Double Load: Women, Work and Feminism on the Canadian Prairie." In *The Prairie West: Historical*

Readings, 2nd ed., edited by R. Douglas Francis and Howard Palmer, 401–23. Edmonton: University of Alberta Press, 1992).

Stuart, Granville. *Forty Years on the Frontier, as seen in the journals and reminiscences of Granville Stuart*, edited by P. C. Philips, vol. 2. Cleveland: A. H. Clark, 1925.

Thomas, Lewis G. *Rancher's Legacy: Alberta Essays*, edited by Patrick E. Dunae. Edmonton: University of Alberta Press, 1986.

Thompson, John Herd. *Forging the Prairie West: The Illustrated History of Canada*. Oxford: Oxford University Press, 1998.

Turner, Frederick Jackson. *The Frontier in American History*. Austin, TX: Holt, Rinehart and Winston, 1962.

Turner, Leland. "Grassland Frontiers: Beef Cattle Agriculture in Queensland and Texas." PhD diss., Texas Tech University, 2008.

Voisey, Paul L. *High River and the Times: An Alberta Community and Its Weekly Newspaper, 1905–1966*. Edmonton: University of Alberta Press, 2004.

———. Vulcan: *The Making of a Prairie Community*. Toronto: University of Toronto Press, 1988.

von Richthofen, Baron Walter. *Cattle Raising on the Plains of North America*. New York: D. Appleton & Co., 1885.

Webb, Walter Prescott. *The Great Plains*. Boston: Ginn, 1931.

West, Elliott. "Families in the West." *Organization of American Historians Magazine of History* 9, no. 1 (Fall 1994): 18–21.

Wetherell, Donald. *Wild Life, Land and People: A Century of Change in Prairie Canada*. Montreal: McGill-Queen's University Press, 2016.

Wheeler, David L. "The Texas Panhandle Drift Fences." *Panhandle-Plains Historical Review* 55 (1982): 25–35.

White, Courtney. *Revolution on the Range: The Rise of a New Ranch in the American West*. Washington, DC: Island Press, 2008.

Williams, H. L. *The Chief of the Cowboys; or the Beauty of the Neutral Ground*. New York: R. Midewitt, [1870].

Wister, Owen. *The Virginian: A Horseman of the Plains*. New York: Macmillan, 1902.

Worster, Donald. *Dust Bowl: The Southern Plains in the 1930s*. Oxford: Oxford University Press, 1979.

INDEX

A

A7 ranch (founded by A. E. Cross, 1885), 3, 21, 24, 133, 207
Abbott, E. C. (Teddy Blue), 121–22, 179, 180
Agricultural Coal and Colonization Company (76 ranch), 10, 13, 53, 79
Alice Springs, Australia, 68
Anderson, Billy, 111
Armour and Company beef merchants, 32
Armstrong, Mrs., 131
Arrowwood, AB, 31
Austin, Fred, 67
Austin, Katherine, 67

B

Baker City, OR, 31, 32
Banff, AB, 103
Bar N ranch, 31, 32
Bar S ranch, 24, 80, 130, 131, 136, 142, 167, 193, 194. *See also* Macleay, Roderick Riddle
Bartch, Chris, 31
Bar U ranch (founded by Fred Stimson, 1881), 1, 24, 10, 11, 53, 113, 131, 206, 207
baseball, 139–40
Bateman, Jessie Louise, 67–68
Beaucook, Bert, 65
Bennett, R. B., 74, 81
Binnie-Clark, Georgina, 158
Black Diamond, AB, 205
Blades, Betty, 62
Blades, Ernie, 62, 85
Blades, Ethel, 62
Blades, Rod, 62, 85
Blades family, 51, 207, 210

Blake, Val, 43, 48, 54, 110, 111, 115, 117–18, 125, 126, 134, 136, 143, 146, 194, 195, 201
Boissevain, MB, 84
Bowers, William L., 105
Breen, David, 2
Brooklyn, ON, 47
Brooks, AB, 3, 26, 73
Bumper, Beatrice, 103
bunkhouse culture, 121–44
Burlington, VT, 38
Burns, Patrick, 72–80, 133

C

Calgary, AB, 3, 67, 81, 107, 113, 140, 143, 144, 145, 159
Capital X ranch, 133
Carter, Sarah, 159
Cartwright family (D ranch founded by Edwin Cartwright and John Thorp circa 1909), 146, 147
Cary, Charlie, 133
Casous, 11
Cayley, AB, 18, 39, 54, 59, 107, 112, 157
CC ranch, 67
Chattaway, Clay, vii, 46, 59, 62, 185, 207, 210
Chattaway, George, 39, 85
Chattaway family, 51, 210
Chicago, IL, 41, 73, 143
Christensen, Carl, 84, 85
Circle Diamond ranch, 10
Circle Three ranch, 53, 73–74, 79, 81
Clay and Robinson, beef agents, 73
Cochrane, Evelyn, 67
Cochrane, Matthew, 133

249

Cochrane ranch (founded by Matthew Cochrane, 1881), 1, 10, 11
Cody, Willliam Frederick (Buffalo Bill), 120
Comrie, Donald, 133
Comrie, Dunk, 66, 110, 133
Comrie, Peter, 133, 147
Comstock, Albert, 131
Connor, Ralph, 13
Copithorne, Susan, 68
Copithorne family (CL ranch founded by Richard Copithorne, 1887), 3
Cory, W. W., 78
country and western culture. *See* bunkhouse culture; *Rocking P Gazette*
cowboys, 11. *See also* bunkhouse culture; dime novels; stampede; stampedes, cattle
Cross, A. E. (Alfred Ernest), 3, 36, 45, 133, 207. *See also* A7 ranch

D

Danville, QC, 9, 18
Daschuk, James, 159
Davis, Walter, 47
Dempsey, Hugh, 159
de Swaan, Abram, 160, 175
Dewdney, Edgar, 163
dime novels, 12–13
Donaghue, J. A., 31
Douglas, Charlie, 36–37
Drumheller ranch, 54
Dryden, Mr., 47
Dunlap, Jim, 11

E

Edmonton, AB, 159
Edward, Prince, 97
Edwards, Henrietta, 158
Ellis, Thomas, 72
Emerson, George, 9, 10, 11, 13–14, 18, 24, 26, 38, 39, 41, 56, 71, 145; ranch, 17, 165

F

Fares, William, 76
Foran, Max, 2
Fort Benton, MT, 140

G

Gale, Hannah, 158
Garceau, Dee, 83
Gardiner, Mr., 147
Gibson, Hoot, 111, 114
Glasgow, Scotland, 46
Glass, Charlie, 39
Gleichen, AB, 31
Gordon, Charles, 76
Gordon, Ironsides and Fares, beef merchants, 26, 41, 76, 77, 79, 209
Gray, James, 2
Graybill, Andrew, 159
Great Falls, MT, 143

H

Hayden, John, 130, 131
Hayes, Mary, 68
Hayes, William, 68
Hendrie, Jim, 43, 99, 110, 123, 125, 127, 131, 194, 197, 201
Henry, William (Billy), 31
Henry ranch, 54
Henson, Clem, 43, 65, 117–18, 123, 146, 195
Hicks, Clara, 84
High River, AB, 9, 14, 26, 30, 43, 59, 63, 71, 73, 106, 113, 145, 191
High River Wheat and Cattle Company, 43, 72, 156
Highwood River, 9, 14, 30
Howe, Sam, 37, 133

I

Inderwick, Mary, 161
Ings, Fred, 140, 164

J

Jenkins, Hugh, 133, 146, 154
Johnson, Everett, 11
Jones, Buck, 111
Jordon, Terry, 2

K

Kerfoot, W. D., 11
Kinnear, Mary, 58, 156
Knox, Charlie, 25, 26
Knupp, Lillian, 41
Koff, P. T. A., 164
Kreps (also spelled Krepps), William (Bill), 54, 60, 111, 117–18, 127, 138, 144, 164, 167, 194, 291
Kyle, Joseph, 79, 80

L

Lane, George, 11, 24, 53, 131, 133, 206
Langdon, AB, 34
Left Hand, Ezra, 165, 168
Lehr, Charlie, 47
Leman, Art, 130, 131
Leman, Mrs., 130
LeMaster, Wick, 117–18, 164, 167
Lethbridge, AB, 32, 45, 143, 159
Liverpool, England, 46,
Livingstone, William (Bill), 54
Lockton, Cecil, 130
London, England, 74
Longfellow, Henry Wadsworth, 170
Longview, AB, 3
Lung, Charlie, 64, 170, 174–75
Lynch, Tom, 9, 11
Lyster, A., 20

M

Maclean, William (Billy), 36
Macleay, Alexander, 15, 17, 75
Macleay, Dorothy, 1, 4, 53, 56, 63, 64, 66, 67, 68, 73, 82, 81, 83, 85, 110, 112, 134, 148, 149, 167, 188, 193, 195, 196, 200, 203, 210; artistic and academic talents, 89–96, 127; blurring gender roles, 64, 65, 150–58; racial values, 165–83
Macleay, Gertrude (née Sturtevant, later Riddle), 38, 63, 188, 200
Macleay, Kenneth, 38, 63
Macleay, Laura (née Sturtevant), vii, 1, 3, 38, 40, 53, 56, 65, 67, 68, 69, 85, 91, 113, 127, 131, 132, 145, 146, 147, 156, 165, 188, 200, 203, 210; artistic talents, 127; full partner on the ranch, 58–64, 71–85, 150; hunting game, 62–63, 91, 103
Macleay, Maxine, 1, 4, 33, 38, 44, 54, 56, 63, 66, 67, 68, 73, 81, 85, 95, 96, 97, 103, 104, 112, 188, 193, 194, 196, 201, 203, 210; artistic and academic talents, 89–91, 127; blurring gender roles, 65, 125, 150–58; racial values, 165–83
Macleay, Roderick Riddle, vii, 1, 3, 17, 18, 19, 20, 33, 34, 39, 58, 59, 62, 64, 91, 101, 107, 113, 117–18, 123, 127, 133, 141, 145, 185, 200, 203, 205, 207, 209, 210; and Bar S ranch, 39, 43, 53, 54, 65, 71–80, 83, 85, 95, 103, 110,123, 125, 127, 131, 167; business dealings with Patrick Burns, 72–80; and Canadian Council of Beef Producers, 45, 193; cattle breeds and quality, 47, 48, 205; cooperation with neighbours, 33, 42, 148; farming practices, 41–56; finances, 71–85, 131; and grain finished beef, 52–53; heads to the West, 9–16; horse business, 29–32; management practices, 41, 53, 54, 188, 193; marketing cattle in the United Kingdom, 46–52; partnership with George Emerson, 38–41; racial values, 164–65; and Red Deer River ranch and grazing lease, 24, 36, 37, 42, 53, 73, 75, 81, 84, 133, 163; and TL ranch, 53, 71–72, 77, 136; and Ware

Index   *251*

ranch, 24–25, 26; and Western Stock Growers' Association, 45, 77, 193; and White Mud River lease, 72–79, 114
Macleod, AB, 36, 113
Macleod, James F., 163, 178
Magnan, André, 2
Manchester, England, 46
Maple Creek, SK, 49
Marshall, C. H., 73
Martin, Miss, 147
Martin, Mrs., 148
Maunsell, Edward, 163
Maunsell, George, 163
McCallum, Mary, 158
McClung, Nellie, 158
McDonal, Jimmy, 136, 137, 139, 167, 194
McDonal, Ralph, 136, 167, 194
McEachran, Duncan, 205
McIntyre ranch (founded by William H. McIntyre, 1894), 3, 45
McKeage, John, 14, 17
McKeage, William (Billy), 14, 17
McKinney, Louise, 158
McKinnon, Tommy, 43, 54, 60, 66, 110, 123, 124, 127, 130, 131, 132, 139, 167, 187, 194
McKinnon family (LK ranch founded by Lachlin McKinnon, 1921), 3
Mclean, A. J., 133
McLean, Carrie, 143
McMillan, R. L., 33, 146
McNaughton, Violet, 158
McNeal, Ida, 158
McNelly, Dan, 147
Medicine Hat, AB, 159
Miles City, MT, 143
Milk River, AB, 34
Miller, Herb, 33
Miller and Dolan, beef agents, 73
Mills, Claude, 191
Mix, Tom, 111
Montreal, QC, 45
Moose Jaw, SK, 73
Morrill, Marvin, 14, 15, 17, 53
Muirhead, Peter, 24, 38, 72

Muirhead School House, 131
Mule Creek Cattle Company, 77, 80
Murphy, Emily, 158
Myres, Sandra, 178

N

Nanton, AB, 21, 67, 146, 163
N Bar N ranch, 10
Nelson, Ted, 48, 194
Neth, Mary, 145
Newport, VT, 58
New York, NY, 53, 74
Nine Six Nine ranch, 13
North West Mounted Police, 36, 163

O

Old Sun, Chief, 164
Omaha, NE, 143
Orvis, Ed, 54, 66, 107–8, 127, 132, 164
Ottawa, ON, 77
Oxley ranch (first managed by J.R. Craig, 1882), 10

P

Palmer, Howard, 159
Parlby, Irene, 158
Patricia, AB, 73
Peddie, George, 54, 66
Pedersen, Johanne, 67
Pekisko Creek ranch, 14
Pendlebury and Maxwell, beef agents, 73
Pincher Creek, AB, 21

Q

Quorn ranch (founded 1885), 205

R

ranching. company approach, 1–2, 9–11, 53, 100, 203–11 (*see also* individual ranches); family approach, 1–3, 17–27, 57–69, 89, 100, 203–11 (*see also* individual ranches; *Rocking P Gazette*; Rod Macleay)
Raynor, Robert, 54, 65, 123, 132, 133, 194
Redcliff, AB, 38
Reeves, Robert (Bob), 63
Regina, SK, 103
Revelstoke, BC, 33
Ribordy, Jack, 66, 117–18, 124
Ricks, Frank, 11
Riddle, Douglas, 14, 15, 17
Riddle, John, 14, 17, 37, 75
Riddle, Margaret, 20, 61
Riddle, Stewart, 39, 54, 56, 65, 72, 75, 107, 110, 117–18, 119, 131, 138, 146, 185, 188, 194, 200
Riddle and Macleay Brothers, 17–40, 206, 207; closing down, 38; and the 1906–07 winter, 33
Riley, Dan, 71, 77
Robertson, Ralph, 117–18, 138
Roblin, MB, 73
Rocking P Gazette, 4–5, 23, 43, 44, 54, 60, 61, 63, 79, 89–211; and conservation, 99–104; and country life movement, 105–6; and country and western culture, 106–20; and family approach to ranching, 185–202; and gender, 150–58; and race, 161–85; and ranch cohesion, 188–201; and religion, 158–61
Rocking P ranch, 39, 41–85, 95, 97, 107, 131, 136; golf course, 98; mixed farming practices, 41–56, 132
Russell, Charles Marion, 127–30, 179–80

S

Saskatoon, SK, 106
Shaddock ranch, 34
Sharpe, Frank, 54, 146
Skrine, Walter, 72
Smith, Char, 143
Smith, R., 54, 110
Smith, Tex, 54, 109–110, 115, 119, 146, 167, 194, 197, 201
Springfield ranch, 21
Stampede. Calgary, 133; Nanton, 134; origin of generally, 133–37
Stampedes, cattle, 113–18
Steele, Sam, 140
Steward, Hon. Charles, 77
St. Paul, Mis., 143
St. Remi Lumber company, 33
Sweet Grass, MT, 32

T

Telocaset, OR, 32
Thompson, Fulton, 71
Thorburn, David, 14
Thurber, Herb, 136, 139, 146
Turkey Track ranch, 10
Turner, Fredrick Jackson, 147

V

Vancouver Prince Rupert Ranching Company, 72
Van Eden, Frank, 54, 111, 115, 116, 123, 125, 131, 164, 167, 194, 199
Voisey, Paul, 2, 3–4, 114, 160, 175, 204
Vulcan, AB, 3–4, 160

W

Waddell, Charlie, 21, 130
Walla Walla, WA, 32
Walrond ranch, 1, 10, 11, 53, 204, 205, 207
Walsh, James Morrow, 178
Walsh, W. L., 74

Walters, C., 54
Walters, Jesse, 194
Walters, Mrs., 188
Ware, Amanda, 25
Ware, John, 11, 24, 25, 38, 131; ranch, 181 (*see also* Macleay, Roderick Riddle)
Ware, Mildred, 24, 25
Ware, Robert Lewis, 25
Watts, Ethel, 91, 95, 100, 101, 127, 131, 147, 154, 167, 168, 175, 177, 188, 193, 194, 196
Weadick, Guy, 133
Weise, Dair, 117–18, 164
Wentworth, Willis, 14, 18
West, Elliott, 57, 67
wheat boom, 30, 43
White, Courtenay, 2
Williams, William (Bill), 117–18, 123
Winnipeg, MB, 26, 30, 76
Winter of 1906–07, 33–37
wolves, 2
Wong, George, 175, 177
Worster, Donald, 2

www.ingramcontent.com/pod-product-compliance
Lightning Source LLC
Chambersburg PA
CBHW041731300426
44115CB00022B/2979